U.S. Fire Administration

Voice Radio Communications Guide for the Fire Service

October 2008

 **FEMA**

U.S. Fire Administration
Mission Statement

We provide National leadership to foster a solid foundation for local fire and emergency services for prevention, preparedness and response.

Dear Members,

Radio communications for the fire service has evolved considerably over the last 60 years. Radios that were once refrigerator-sized monsters are now small enough to fit in the palm of the hand.

It used to be that only the company officer was permitted to use a radio. Today, radios are a critical safety tool that must be in the hands of every fire fighter at every emergency scene. As advances in radio communication technology occur, it's important to make sure that radios remain an effective and reliable means of communication.

Specifically, new technology for radio communication systems must meet the unique demands of the job of fire fighting. Fire fighters must be able to communicate in cold and hot temperature extremes, in wet and humid atmospheres full of combustion byproducts and dust, while under or above ground, inside and below buildings and in rubble piles. Other environmental challenges include loud noise from apparatus, warning devices, tools and the fire itself. Any new radio communication system must take these factors into consideration.

The IAFF has made it a priority to ensure that everyone goes home safe at the end of each shift. Because radios are one of most important pieces of safety equipment, we expect that any new communications system will be effective, safe, reliable and simple to use.

I urge every IAFF affiliate to be involved early on in the process of developing a new radio communication system in their jurisdiction to make sure that the funding, staffing, training, testing, trouble-shooting and implementation meet the standards and requirements for fire fighters to respond safely and effectively.

This Manual is designed to help affiliate leaders and members understand new communication and radio system issues in order to remain informed players in the process.

An effective communications system requires proper planning at the front end in order to prevent problems later, and there is no one better to participate in the process than fire fighters.

Stay safe,
Harold A. Schaitberger
IAFF General President

ACKNOWLEDGEMENT

The United States Fire Administration (USFA) is committed to using all means possible for reducing the incidence of injuries and deaths to firefighters. One of these means is to partner with organizations that share this same admirable goal. One such organization is the International Association of Fire Fighters (IAFF). As a labor union, the IAFF has been deeply committed to improving the safety of its members and all firefighters as a whole. This is why the USFA was pleased to work with the IAFF through a cooperative agreement to develop this *Voice Radio Communications Guide for the Fire Service*. The USFA gratefully acknowledges the following leaders of the IAFF for their willingness to partner on this project:

General President
Harold A Schaitberger

General Secretary-Treasurer
Vincent J. Bollon

Assistant to the General President
Occupational Health, Safety & Medicine
Richard M. Duffy

International Association of Fire Fighters, AFL-CIO, CLC
Division of Occupational Health, Safety and Medicine
1750 New York Avenue, NW
Washington, DC 20006
(202) 737-8484
(202) 737-8418 (FAX)
www.iaff.org

The IAFF also would like to thank Leif Anderson, Deputy Chief, Phoenix Fire Department; Jim Brinkley, IAFF Director of Occupational Health and Safety; Joseph Brooks, Radio Supervisor, Boston Fire Department; Missy Hannan, Senior Graphic Designer, International Fire Service Training Association (IFSTA)/Fire Protection Publications, Oklahoma State University; Tim Hill, Captain, Phoenix Fire Department and President of the Professional Fire Fighters of Arizona; Christopher Lombard, Lieutenant, Seattle Fire Department; Andy MacFarlane, Phoenix, Arizona; Brian Moore, Captain, Phoenix Fire Department and Director of Member Benefits, IAFF Local 493; Kevin Roche, Assistant Fire Marshal, Phoenix Fire Department; Mike Wieder, Assistant Director, IFSTA/Fire Protection Publications, Oklahoma State University; and Mike Worrell, Captain, Phoenix Fire Department, for their efforts in developing this report.

TABLE OF CONTENTS

SECTION 1
INTRODUCTION

Purpose

The past few decades have seen major advancements in the communications industry. Portable communications devices have gone from being used mainly in public safety and business applications to a situation where they are in every home and in the hands of almost every American man, woman, and child. As users are added, there is more stress on the system; there is only so much room on the radio spectrum. The communications industry and the government have responded by making changes to the system that mandate additional efficiency.

These advancements have improved radio frequency spectrum efficiency, but also have added complexity to the expansion of existing systems and the design of new systems. Some of these advances in technology are mandated by the Federal Communications Commission (FCC), while others are optional. The costs and operational effects of these changes are significant. Navigating through the complex technological and legal options of public safety communications today led to the development of this guide to assist the fire service in the decisionmaking process.

Why the Fire Service is Different

The life safety of both firefighters and citizens depends on reliable, functional communication tools that work in the harshest and most hostile of environments. Firefighters operate in extreme environments that are markedly different from those of any other radio users. Firefighters operate lying on the floor; in zero visibility, high heat, high moisture, and wearing self-contained breathing apparatus (SCBA) facepieces that distort the voice. They are challenged further by bulky safety equipment, particularly gloves, that eliminate the manual dexterity required to operate portable radio controls. Firefighters operate inside structures of varying sizes and construction types. The size and construction type of the building have a direct impact on the ability of a radio wave to penetrate the structure. All of these factors must be considered in order to communicate in a safe and effective manner on the fireground.

Radio system manufacturers have designed and developed radio systems that meet the needs of the majority of users in the marketplace. The fire service is a small part of the public safety communications market and an even smaller part of the overall communications market. This has resulted in one-size-fits-all public safety radio systems that do not always meet the needs of the fire service as a whole or those of a specific department. For a number of reasons, the fire service is unique among public safety and other municipal communications users. A large percentage of radio communications by most municipal or government users is done from vehicle-mounted mobile radios. Public safety radio users, law enforcement, and the fire service, use vehicle-mounted and portable radios. For much of the portable radio work by law enforcement personnel, the officer is outside on the street, in an upright position, with good visibility. The officer occasionally will go inside a building and communicate. This is in sharp contrast to the environments that firefighters face on a daily basis.

In many instances, radio systems perform better for law enforcement than for the fire service. The main reason for this is that law enforcement operates differently than the fire service. In the law enforcement operating model, the officers are in a deployed state outside patrolling the streets. When an incident occurs, the dispatch center notifies the patrol officers of an incident, and an officer or officers respond to the call. Once officers arrive onscene they may be operating as a single resource and only require communication with the dispatcher; at other times they may be operating with multiple responders, but the dispatcher remains the focal point of communications. Officers in these situations are wearing standard patrol attire and have good visibility.

The fire service operates in a staged state with resources located in fire stations. Calls are dispatched to specific units based on their location in relation to the incident. When more than one unit responds to an incident, an onscene Command structure is established to coordinate fire attack, provide safety and accountability, and manage resources. The units assigned to these incidents work for the local Incident Commander (IC), who is the focal point of communications on the fireground. The dispatch center assumes a support role and simultaneously documents specific fireground events, handles requests for additional resources, and may record fireground tactical radio traffic.

When comparing law enforcement to the fire service and other public safety some major differences are apparent.

Fire Service	Law Enforcement
Majority of incidents in buildings	Majority of incidents on street
Contaminated breathing atmosphere requiring SCBA	Safe breathing atmosphere
Often operate in a prone position	Upright position
High temperatures	Normal temperatures
Poor voice quality to radio	Good voice quality to radio
High background noise on incident scenes	Normal to high background noise
Poor to zero visibility	Good visibility
Poor to no manual dexterity	Good manual dexterity
Local Command structure coordination Localized communications	Dispatch center coordination Wide area communications

SECTION 2
BASIC RADIO COMMUNICATION TECHNOLOGY

When talking about fire department communications systems usually we are talking about what are traditionally called **land mobile radio systems.**

It is important for firefighters and fire officers to have a basic knowledge of radio system technologies to help them during the design, procurement, or use of the radio system. By having this basic understanding, you will be able to participate effectively in critical discussions with technical staff, consultants, and manufacturers to get the safest, most effective voice communications system for your firefighters, Command Staff, and community.

Most radio system users do not need a detailed understanding of the technology behind the systems they use. However, such knowledge is important for those involved in procuring the systems, in developing procedures for the use of the systems, and in training field users to have a more comprehensive understanding of their operation. All technologies have strengths and weaknesses, and understanding those characteristics is important in making decisions related to the technologies. No matter what a salesperson will tell you during the procurement process, no system is without risk and all have had users who were not satisfied with some aspect of the system. The key is in understanding the technology enough to ask questions, understand the answers, and make a successful evaluation.

Radio Spectrum

Radio communications are possible because of electromagnetic waves. There are many types of electromagnetic waves, such as heat, light, and radio energy waves. The difference between these types of waves is their frequency and their wavelength. The frequency of the wave is its rate of oscillation. One oscillation cycle per second is called one hertz (Hz). The types of electromagnetic energy can be described by a diagram showing the types as the frequency of the waves increase.

Figure 1 – The Electromagnetic Spectrum.

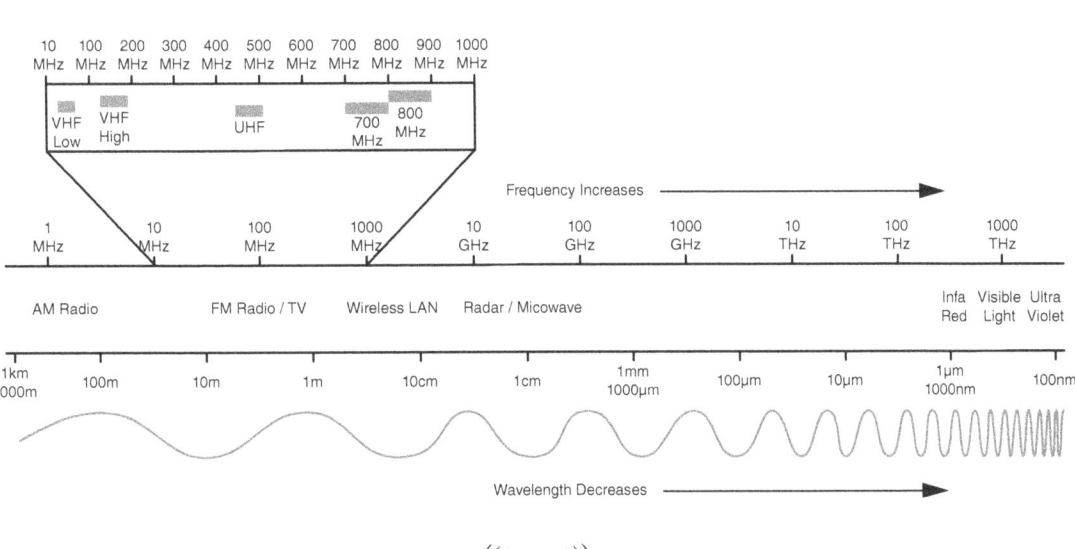

When describing the frequencies used by common radio systems, we use the metric system to quantify the magnitude of the frequency. A typical frequency used in fire department radio systems is 154,280,000 Hz. This is a frequency designated by the FCC as a mutual-aid radio channel. Dividing the frequency by the metric system prefix mega, equal to 1,000,000, this becomes 154.280 megahertz or MHz.

Land mobile radio systems are allowed to operate in portions of the radio spectrum under rules prescribed by the FCC. These portions of the spectrum are called bands, and land mobile radio systems typically operate with frequencies in the 30 MHz (VHF low), 150 MHz (VHF high), 450 MHz (UHF), 700 MHz, and 800 MHz bands.

The wavelength is the distance between two crests of the wave. The frequency and wavelength are inversely related so that, as the frequency of the wave increases, the wavelength decreases. The length of a radio antenna is related to the wavelength with which the antenna is designed to operate. In general, the higher the frequency of the waves used by the radio, the shorter the antenna on the radio.

Figure 2 – Electromagnetic Wave.

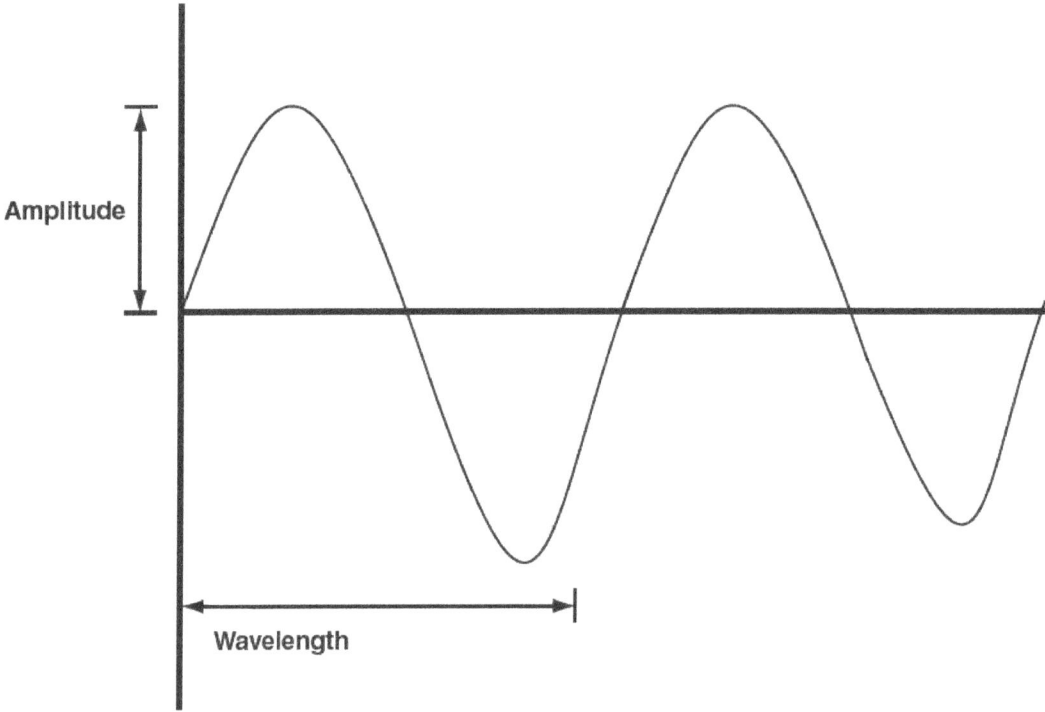

Channel Bandwidth

The radio spectrum is divided into channels. Each radio channel is designated by a frequency number that designates the center of the channel, with half of the bandwidth located on each side of the center.

Radio channel bandwidth is the amount of radio spectrum used by the signal transmitted by a radio. The greater the bandwidth, the more information can be carried by the signal in the channel. Minimum channel bandwidth typically is limited by the state of technology, and the bandwidth required to carry a given amount of information has decreased by several times over the past 50 years. However, there is a theoretical limit below which the bandwidth cannot be decreased. In addition, the actual width of a channel often is slightly greater than the minimum width, to provide some space on each side of the signal for interference protection from adjacent channels. For the purposes of radio licensing, the FCC sets the maximum and minimum bandwidth for channels in each frequency band.

The bandwidth of channels typically used in land mobile radio is measured in thousands of hertz, or kilohertz, abbreviated kHz. In an effort to place more communications activity within a limited radio spectrum, permitted bandwidth has been decreasing. Under older licensing rules, some of which are still in effect, typical channel bandwidths were 25 kHz. Newer rules require bandwidths of 12.5 kHz.

Figure 3 – Channel Bandwidth.

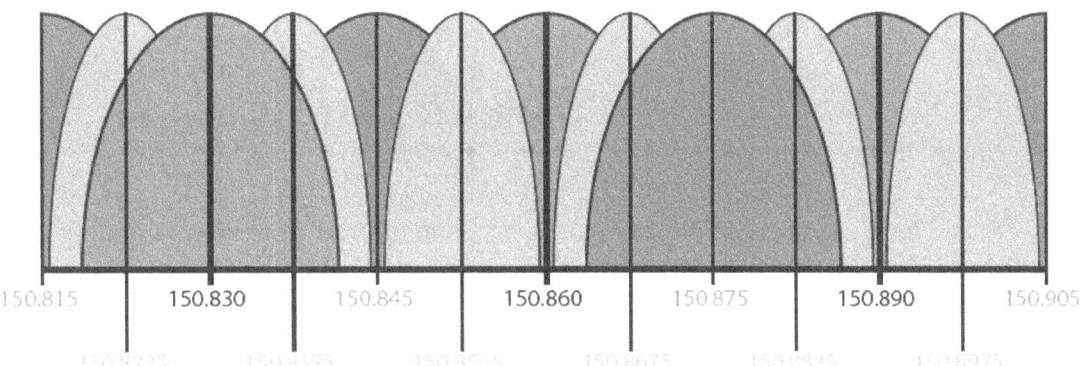

150.815   150.830   150.845   150.860   150.875   150.890   150.905

150.8225   150.8375   150.8525   150.8675   150.8825   150.8975

▦ Existing 25 kHz bandwidth channels spaced at 30 kHZ intervals.
▦ Existing 25 kHz bandwidth channels spaced 15 kHZ from the original channels.
▦ New 12.5 kHz bandwidth channels at 7.5 kHZ spacing from existing channels.

Radio Wave Propagation

To send a radio signal from a transmitter to a receiver, the transmitter generates electromagnetic energy and sends that energy through a transmission line to an antenna. The antenna converts the energy into electromagnetic radio waves that travel at the speed of light outward from the antenna. If another antenna is located in the path of the waves, it can convert the waves back into energy and send that energy through a transmission line to a receiver.

Figure 4 – Electromagnetic Signal Radiation.

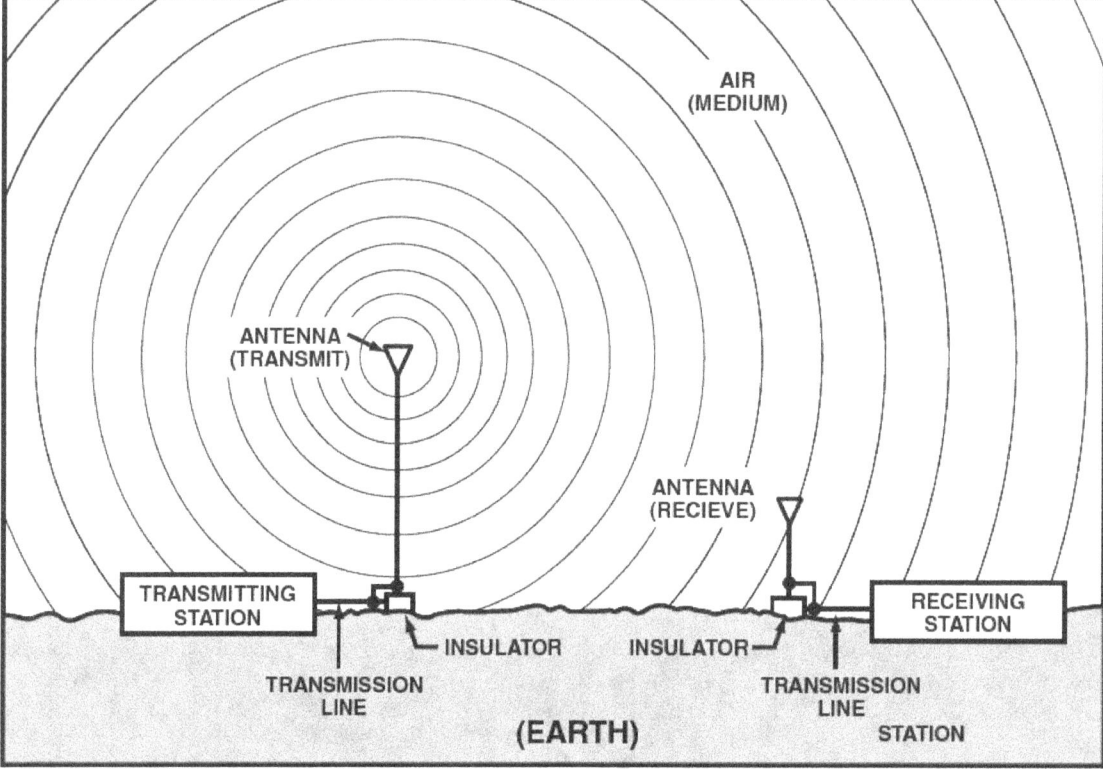

Radio signals emitted from an antenna travel both a direct path to the receiving antenna, and a path reflected from the ground or other obstacles. This reflection causes the wave to travel a longer distance than the direct wave, as shown in **Figure 5.**

Figure 5 – Signal Paths.

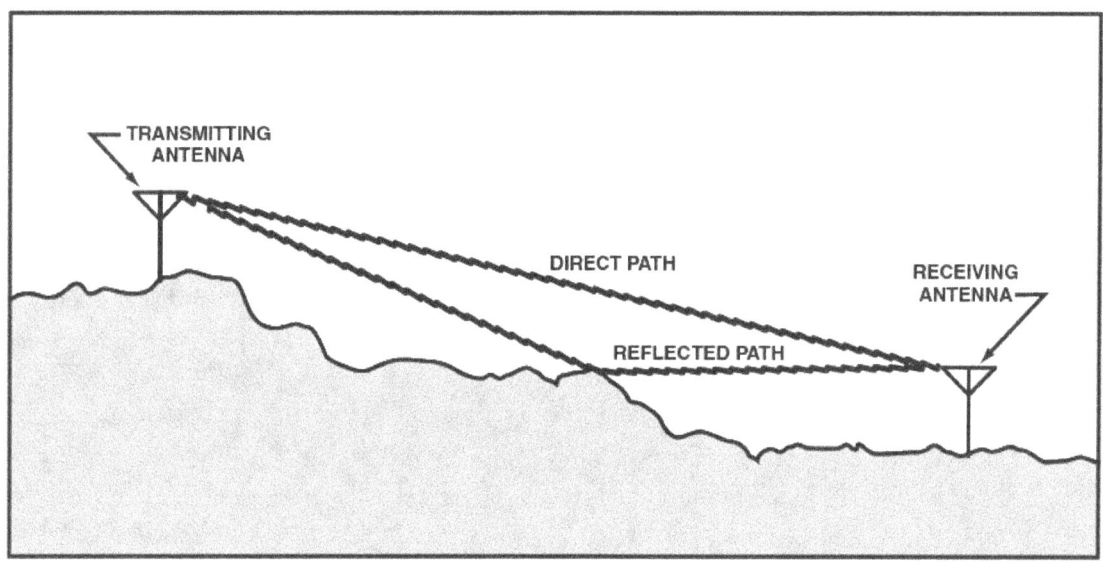

The waves traveling over the reflected path then interfere with the direct waves, causing an effect know as multipath interference. Multipath interference causes a variation in the signal level at the receiver. The signal may be higher or lower than the direct signal depending on the position of the receiver's antenna. As the antenna is moved around, the signal varies, and the user hears a signal that goes from strong and clear to weak and noisy.

Radio waves can travel through some materials, such as glass or thin wood, but the strength is reduced due to absorption as they travel through. Materials such as metal and earth completely block the waves due to their composition and density. In addition, some materials will reflect radio waves, effectively blocking the signal to the other side.

Because buildings are built from many types of materials, the radio waves can be passed through some, be reflected by some, and be absorbed by others. This, along with the complex interior design of a building, creates a very complex environment for radio communications inside a building.

Figure 6 – Terrain Blocking.

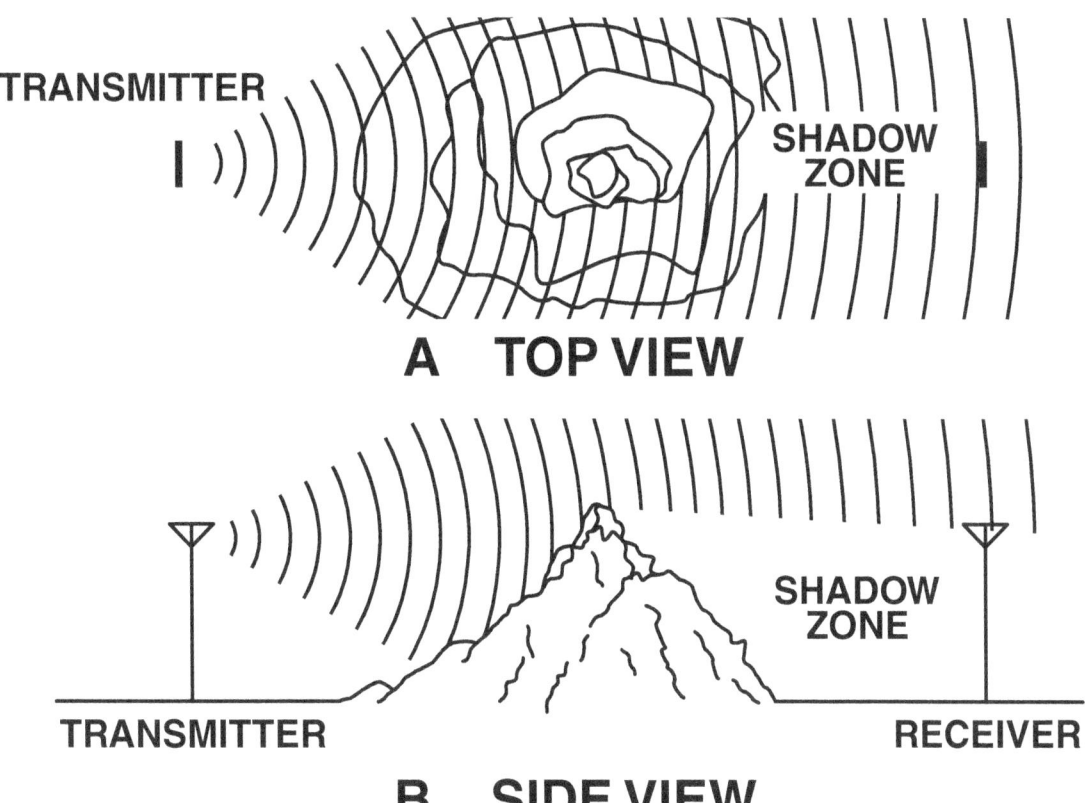

Interference

Radio frequency interference can be either natural or manmade. Interference from internal noise occurs naturally in all electronic equipment due to the nature of the electronic circuit itself. Manufacturers take this into account during equipment design, and obtaining a low-noise design is not particularly difficult. In addition, natural noise is produced by sunspot activity, cosmic activity, and lightning storms. This noise usually is of small magnitude and not significant for most land mobile radio communications. However, the VHF low band is affected significantly by severe sunspot activity, sometimes to the point of completely prohibiting communications.

More significant to radio communications systems is the interference produced by manmade sources. Vehicle ignitions, electric motors, high-voltage transmission lines, computers, and other equipment with microprocessors also emit radio signals that can interfere with public safety radios.

In general, manmade interference decreases with an increase in frequency. The UHF band and, initially, the 800 MHz band are much less susceptible to manmade interference than the VHF low and high bands. When systems are not subject to significant interference, they are said to be "noise limited," in contrast to "interference limited." The large number of transmitters used by cellular telephone companies has created intense interference in the 800 MHz band.

Although the separation of the channels allocated to cellular companies has reduced this interference, communications problems still can occur when a user is operating close to a cellular transmission facility. This type of interference is particularly a problem when the user is located near a cellular facility and the user's radio system site is located much further away. This creates a situation called near-far interference. The user's system signal strength is low, and the cellular signal is high, keeping the user's radio from receiving the desired signal. The 800 MHz band always was regarded as the "cleanest" band with respect to manmade interference, and systems initially were noise limited. However all systems in the band now must be designed for maximum interference from nearby transmitters, requiring more transmitter locations and higher power creating more costly systems.

Interference from cellular transmitters is illustrated in **Figure 7.** The blue area in the center is the public safety transmitter and in the center of the grey areas are the cellular transmitters.

Intermodulation interference is caused directly by the mixing of two or more radio signals. The mixing most commonly occurs inside the receiver or transmitter of a radio. This mixing can create a third signal that is radiated from the antenna out to other radios. The mixing also can occur outside a radio in the transmission line or through rusty tower bolts or guy wires. Intermodulation can be difficult to identify, due to the large number of frequencies that may be present at large communications sites.

Figure 7 – Gray Areas are Near – Far Interference Holes.

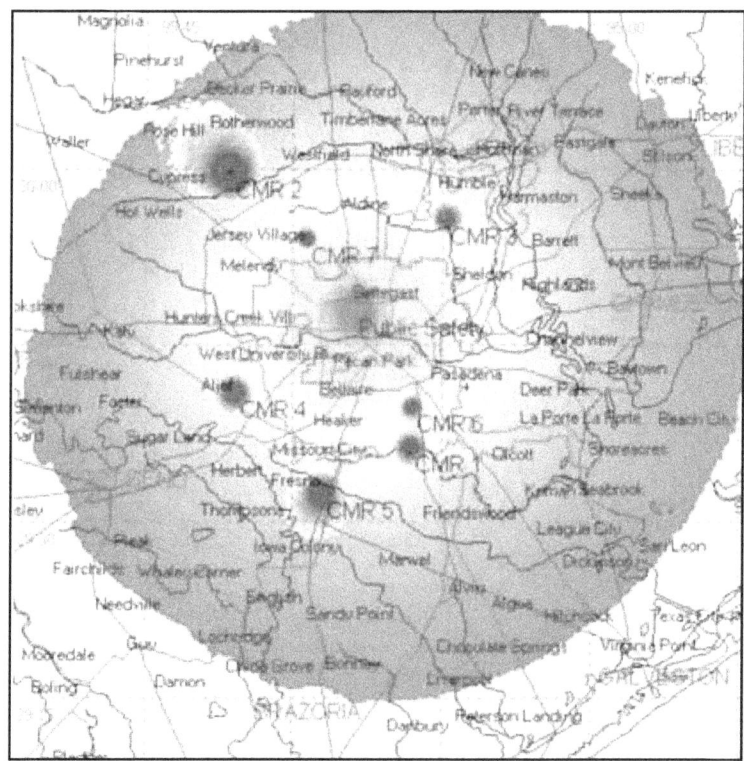

Receiver desensitization interference, also called receiver overload, is caused by nearby high-level transmitter signals that overload the initial parts of the radio's receiver. This overload prevents the receiver from detecting the weaker desired signals, making the receiver nonfunctional. Receiver desensitization occurs near high-power radio sites, such as television and radio stations, and also can occur in poorly designed repeater systems where the transmit and receive frequencies are too close in frequency.

Several things can be done to reduce or eliminate interference. The first is the use of high-quality radio equipment. High-quality equipment has better transmitter and receiver performance that minimizes interference and reduces its effects. The use of receiver multicouplers, transmitter combiners, and repeater duplexers reduces the possibility of intermodulation and receiver overload by filtering the transmitter and receiver signals to ensure only those signals actually used by the system are passed through.

Radio system designers can reduce the possibility of their systems causing interference by selecting appropriate designs. By selecting the appropriate antenna and adjusting transmitter power levels, the system can minimize interference with other users of the same frequency. This allows more efficient use of the available radio spectrum and keeps more resources available for all users.

What Affects System Coverage?

The coverage of a radio communications system generally is described as the useful area where the system can be used reliably. Many factors affect coverage, including the radio power output, antenna height and type, and transmission line losses. However, the factor that most influences coverage is the height of the antenna above the surrounding ground and structures. By locating the antenna on a tower or mountain top, the system designer provides a more direct path from the transmitter to the receiver. In the case of one radio user transmitting directly to another radio user, having the radio antenna as high as feasible (hand held at shoulder height) significantly improves coverage.

Antennas have three major properties: operating frequency, polarization, and radiation pattern. In general, these properties apply whether the antenna is used for transmitting or receiving. The **operating frequency** of an antenna is the frequency at which the antenna acts as specified by its manufacturer. The antenna may operate outside its design frequency, but the performance of the antenna will be reduced.

In land mobile radio systems like those used by public safety, most antennas are **vertically polarized**. You can see evidence of this with the wire antennas mounted on the roofs of vehicles. Like car antennas designed for frequency modulation (FM) broadcast radio, they stick up vertically from the surface of the vehicle.

The **radiation pattern** of the antenna is the shape of the relative strength of the electromagnetic signal emitted by the antenna, and this depends on the shape of the antenna. The radiation pattern can be adjusted through antenna selection to provide coverage where desired and to minimize coverage (and, in turn, interference) in undesired directions.

Fixed-Site Antennas

Fixed-site antennas are mounted on towers or buildings to provide the dispatch or repeater coverage throughout the service area. The antennas used must be designed to operate in the system's frequency band and, for best power coupling, should have a center frequency as close as possible to the actual operating frequency.

The radiation pattern for the antenna should be selected to provide a signal in the desired sections of the coverage area, and have minimal coverage outside the desired coverage area. This will help ensure that the system is not interfering with other systems unnecessarily. The most basic practical antennas are omni-directional, and have approximately equal coverage for 360 degrees around the antenna.

Figure 8 – Antenna Tower and Antennas.

However, as shown in **Figure 9**, the antenna pattern is more like a slightly flattened donut. This causes an area immediately under the antenna to have lower signal strength, and less coverage, than farther away from the antenna.

Directional antennas are used to direct the signal toward the users and away from unwanted areas. The antenna is said to have gain over an omnidirectional antenna in the direction of highest signal. **Figure 10** shows a directional antenna called a Yagi, along with its radiation pattern looking down on the antenna. The pattern shows a stronger signal from the front of the antenna and a weaker signal from the back. The signal strength protrusions behind the main signal are called lobes and, in most cases, antenna designers strive to minimize this unintended signal.

Figure 9 – Omnidirectional Antenna Pattern.

Figure 10 – Directional Antenna and Pattern.

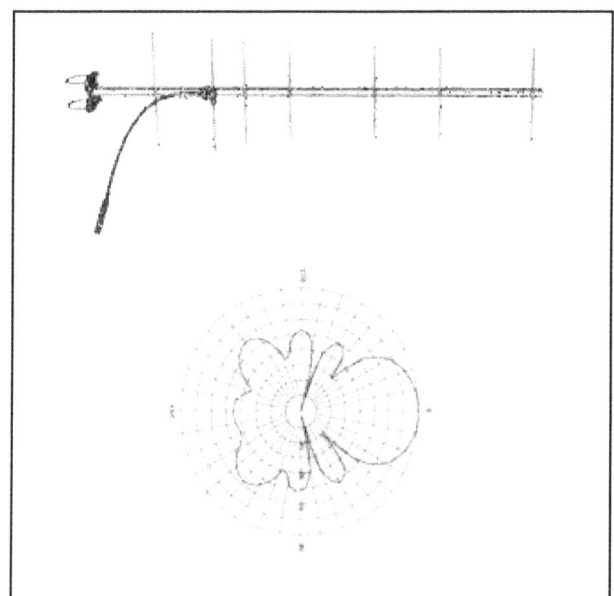

Figure 11 – Down Tilt.

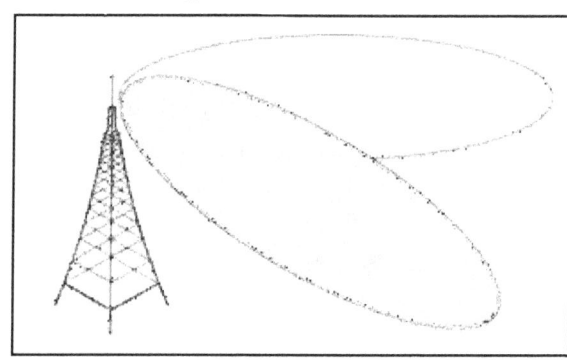

When an antenna is located on top of a mountain or tall building, the coverage loss created by the "hole" in the radiation donut may have a significant impact on coverage in the area immediately around the antenna. To compensate for this, a directional antenna can be tilted slightly to direct more of the signal downward, as shown in **Figure 11**. This tilting is known as mechanical down-tilt and increases the energy immediately below the antenna while reducing the maximum distance the signal will travel. Unfortunately, when using an omni-directional antenna, tilting the antenna down in one direction results in tilting the pattern up on the opposite side of the antenna. For this reason, special antennas with electrical down-tilt are used when omnidirectional coverage is required, such as on a tall building in the center of the coverage area.

Mobile and Portable Antennas

In general, all mobile and portable radio antennas are omnidirectional to provide coverage 360 degrees around the radio user.

Vehicle antennas should be mounted so that they are not obstructed by equipment mounted on the top of the vehicle. Light bars, air conditioning units, and master-stream appliances are some typical obstructions found on fire service vehicles. Some obstructions, such as aerial ladders on truck companies, cannot be avoided, and the designer must select the best compromise location.

Vehicle antennas mounted on the roof of fire apparatus can be damaged by overhead doors, trees, and other obstructions. Ruggedized low-profile antennas often are a better choice, even if they have a lower gain than a normal whip antenna. A properly mounted intact antenna with a lower gain is much better than a damaged antenna of any type.

Portable antennas usually are provided by the portable radio manufacturer and are matched to the radio. In some cases alternative antennas can be selected for the radio to overcome specific user conditions.

When a portable radio is worn at waist level, such as with a belt clip or holster, the user's body absorbs some of the signal transmitted or received by the radio. In addition, the antenna is at a much lower level than if the user were holding the radio to his or her face for transmitting.

Figure 12 – Relative Signal Levels.

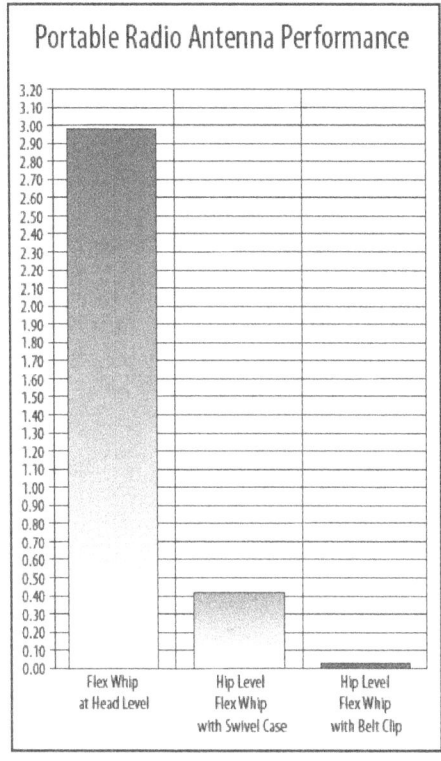

Since the radio system is designed for use with the antenna oriented vertically, the performance of the radio is reduced when the antenna is horizontal. This is particularly important for firefighters, since the radio they use may become oriented horizontally when they are crawling low inside a structure fire.

Figure 13 – Radio Antenna Placement.

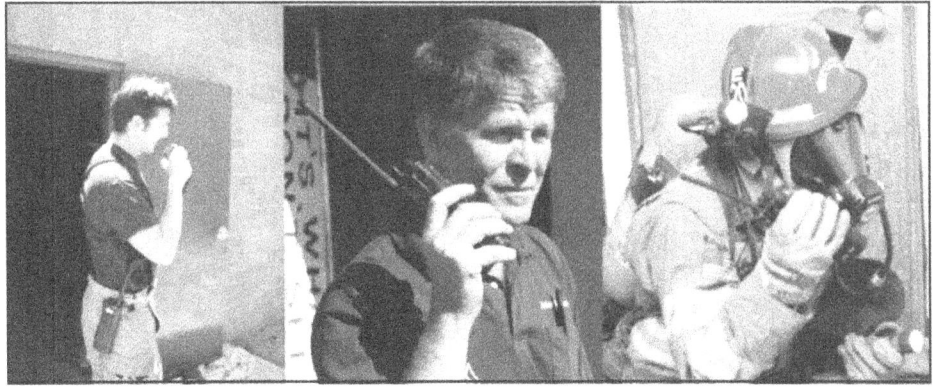

Summary

Radio communication takes place using electromagnetic waves that travel from the transmitter to the receiver. These waves can be reflected or absorbed by materials such as buildings, the earth, or trees, reducing the strength of the wave when it reaches the receiving antenna. Elevating the transmitting or receiving antenna will reduce the likelihood of the wave being affected by buildings or trees, because the path to the receiver will be more direct.

Interference from undesired radio waves is always a possibility in a radio system. The potential for natural interference decreases as the frequency band increases, but manmade interference is very high in the 800 MHz band due to the proximity of cellular and other non-public safety communications systems. This interference can make it difficult to communicate effectively in the presence of the interference.

When designing radio communications systems, the designers must take into account the presence of reflecting or absorbing materials and interference. This may require constructing taller towers to support the antennas or increasing the power of the transmitters to overcome the loss of signal strength and interference. The system design must take into account local terrain, trees, buildings, and the density of interference-generating sources.

SECTION 3
RADIOS AND RADIO SYSTEMS

Several different types of radios are used in the fire service. These radios can be classified as mobile, portable, or fixed; analog or digital; and direct, repeated, or trunked radios. In this section we discuss the operation of these types of radios, and the features, benefits, and problems associated with their use in the fire service.

Mobile radios are designed to be mounted in vehicles and get their power from the vehicle's electrical system. They can be of either a one- or two-piece design, with the radio itself separated from the controls. An external antenna is connected to the radio and permanently mounted to the vehicle. Mobile radios usually have better performance than portable radios, including better receivers and more powerful transmitters. Mobile radios used in trunked radio systems may or may not have more powerful transmitters because the systems are designed for portable use, reducing the need for high-powered transmitters.

Portable radios are hand-held radios powered by rechargeable or replaceable battery packs. They usually have an external rubber antenna attached to the top of the radio.

Mobile and portable radios have similar controls to perform their essential functions. These include things such as changing channels, adjusting the speaker volume, and transmitting. The common names for these controls are the channel (or talkgroup) selector, volume adjustment, and push-to-talk (PTT) switch. Some radios, particularly those intended for fire and police use, will have an orange or red EMERGENCY button. This button may be programmed to indicate to the radio system, and to other users, that a user has an emergency. Older radios may have a squelch adjustment knob, but most modern radios have internal control settings or adaptive squelch so that a squelch adjustment knob is no longer necessary.

Base station radios are located at fixed locations, and usually are powered by AC utility power. Base stations generally are higher in performance than mobile and portable radios, with higher-powered and more stable transmitters and more sensitive and interference-resistant receivers. Some fire departments equip fire stations with base station radios to provide enhanced coverage throughout their service area and to provide backup communications in the event of a primary communications system failure.

Repeaters are similar to base stations, but they can transmit and receive at the same time, retransmitting the signal received by the receiver. Repeaters are used to extend the coverage of portable or mobile radios.

Radio console equipment is used by dispatchers to control base station radios and repeaters and allow the dispatcher to receive and transmit on one or more radios simultaneously. The consoles typically have individual volume and transmit controls for each radio as well as a master volume and transmit control. Headsets can be connected to the consoles along with footswitches, allowing dispatchers to operate the console hands-free so they can operate computer equipment simultaneously.

Analog Radios

The human voice is an analog signal; it is continuously varying in frequency and level. Analog radios have been in use since the invention of voice radio in the early 1900s. The type of analog radio used today was invented in the 1930s to improve on the older radio's poor immunity to noise. These radio systems use frequency modulation (FM) to modulate the transmitted signal with the user's voice. The main advantage of FM over older radio system types is that FM radios tend to reject (interfering) signals that are weaker than the desired signal.

Analog FM radios operate by causing the transmitting frequency of the radio to change directly with the microphone audio. Initially the signal is filtered to remove any frequencies above human voice, but no other changes are made to the signal. **Figure 14(A)** shows an example signal from the microphone, and **Figure 14(B)** shows the resulting change in frequency of the transmitted signal.

Figure 14 – Frequency Modulation.

FM radios constantly have a signal at the output of the receiver, and a squelch circuit is used to mute the output of the radio receiver when no desirable signal is present. Noise squelch circuits mute the output as long as only squelch noise is present. Most older radios feature adjustable squelch level controls, allowing the user to make the radio less sensitive if there is interference. However, most new radios have improved receiver performance and have fixed squelch levels or levels that are adjustable only by radio technicians.

To further reduce received noise and interference, well-designed analog radio systems use tone-coded squelch (TCS) or digital coded squelch. TCS is also known by its Motorola trademark Private Line™ or PL and by its GE (now M/A-COM) trademark Channel Guard™.

TCS systems mix a subaudible tone with the audio from the microphone and transmit the resulting signal. When a radio receives a signal with tone-coded squelch, the TCS decoder attempts to match the tone present in the received signal with the desired tone. If the correct tone is present, the receiver is unsquelched and audio is routed to the speaker.

Digital Radios

To improve audio quality and spectrum efficiency, radio manufacturers introduced digital radios. This was a necessary move based on the FCC requirements to continue narrow banding. Since being introduced, digital modulation has been touted as having more user features, better audio clarity, and higher spectrum efficiency than analog modulation.

In the digital world when a user speaks into the microphone the radio samples the speech and assigns the sample a digital value. A **vocoder** (voice coder) or **codec** (coder/decoder) in the radio performs the function of converting analog voice to a digital data packet. The digital data packet can vary in the number of bits. The higher number of bits in the data packet, the higher the level of precision. Numerous samples are taken each second to reproduce the source audio. The higher sample rate per second and number of bits per sample result in increased audio quality. For example, compact disc (CD)-quality audio samples 44,100 times per second and the number of graduations in the sample is 65,536. The use of digital audio was expected to reduce static and increase the range of radios in weak signal conditions.

Specification	CD Audio	DVD Audio	Standard Public Safety Digital (P25) Audio
Sampling Rate	44.1 kHz	192 kHz	50 Hz
Samples Per Second	44,100	192,000	50

Digital Audio Processing

A vocoder in a digital radio converts analog voice to a digital interpretation from an audio sample. Digital radios, unlike CD or DVD audio, have very limited data rates. Even cell phones have higher data rates than a digital radio. Because of limited data rates, digital radio audio is sampled at a much lower rate with less precision. Designers of the portable radio vocoders felt the radios did not need the same level of precision as CD-quality audio, since reproduction of human speech was the goal.

This is a basic explanation of how analog voice is processed by the radio.

Transmitting radio:

1. The user speaks into microphone.
2. The audio is sampled and converted to a digital interpretation by an analog to digital converter (A/D converter).
3. The vocoder converts the digitized speech into digital data.
4. The modulator modulates the radio frequency (RF) with the digital data.
5. The modulated RF signal is boosted in power by transmitter amplifier.
6. The signal is transmitted from the radio antenna.

Receiving radio:

1. The modulated RF is received by antenna.
2. The received RF signal is boosted to a useable level by the receive amplifier.
3. The signal is demodulated by a demodulator. This removes the RF component of the signal leaving the digital data component.
4. Digital data is decoded by the vocoder into digitized speech.
5. Speech data is converted to an analog signal by a digital to analog converter (D/A converter).
6. Analog is sent to the speaker.

Figure 15 - Digital Radio.

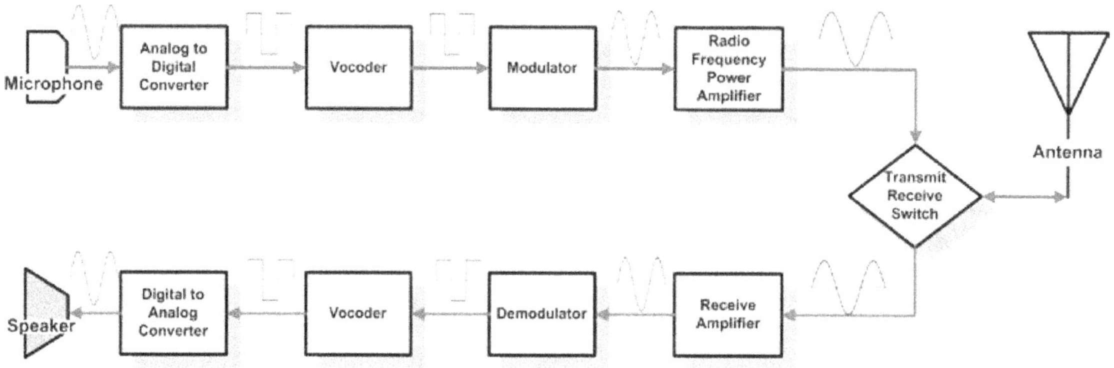

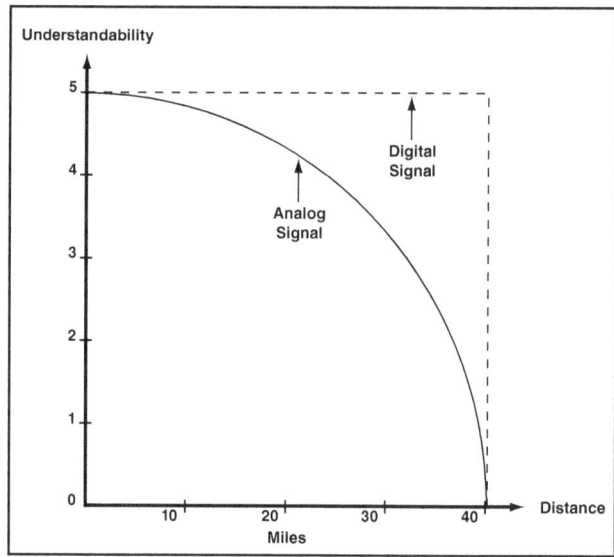

Analog versus Digital Signal Variations With Signal Strength

As the radio user travels further from the transmitting radio, the signal strength decreases. The signal strength directly affects the ability of the radio to reproduce intelligible audio.

In an analog system, the clarity and intelligibility of the transmission, as received by the user, decrease directly as the signal level decreases. The noise (static) in the signal progressively increases in strength, while the desired signal decreases, until the transmitting user cannot be heard over the noise.

When a digital user transmits to a receiver, the transmitted signal decreases just as the analog signal does. However, the error correction in the digital transmission contains extra information that allows the audio information to be heard even with a large decrease in signal level. As the receiver travels further from the transmitter, the signal level decreases to the point where the error correction cannot correct all errors in the signal. When this point is reached, the receiving users will hear some distortion in the signal and may hear some strange nonspeech noises. These strange nonspeech noises are sometimes called "Ewoking" after the language spoken by the Ewok characters in the movie "Star Wars".[1] Once this point is reached, a small reduction in signal level will cause the number of errors to exceed the ability of the system to compensate, and all audio will be lost.

The problem this causes is that the radio signal goes from usable to unusable with little or no indication that this is about to occur. **Figure 16** shows that, with an analog radio system, the signal slowly gets noisier, giving the user hints that the signal is getting weaker. This behavior adds to the situational awareness of the user and allows him/her to make decisions about the environment. Although digital radios provide a larger range of usable signal levels, the lack of advance indication of signal level decrease allows users to get closer to complete loss of communication without any advance warning. Some radio manufacturers have indicated they would develop solutions to this problem; however there are no solutions implemented in current production radios.

Figure 16 - Analog versus Digital Signal.

APCO P25

The Association of Public Safety Communications Officers (APCO), representing the public safety technical community and the Telecommunications Industry Association (TIA), recognized there would be a requirement to move to digital technology. This provided an opportunity to develop an open standard that would allow different manufacturers to build equipment that could operate together. The goal was to introduce competition into the market, help control costs, and provide a technology platform for improved interoperability.

Up until the development of this standard, each manufacturer had proprietary digital radios that could interoperate only with like radios. Working with the TIA, APCO coordinated the work of manufacturers to develop the **P25 Standard** for digital radios. Modern public safety digital radios use this standard. P25 is the national standard for public safety digital radios, but also is backward compatible for analog use. This standard was developed to allow radios from multiple manufacturers to communicate directly using a common digital language, define standards for trunked radio systems to allow multiple manufacturers to operate on a common platform, and to provide a roadmap for future features and capabilities.

What P25 is Not

P25 does not address any operational or interoperability needs. P25 does not provide a fire department with interoperability unless it is planned for. A lone agency on P25 is no more interoperable than being on a UHF system trying to interoperate with a department on VHF. P25 only provides manufacturers with a common digital language for the radios and system infrastructures.

P25 system standards also were meant to allow radios of different manufacturers to operate on any other P25-trunked radio system. This has not been the case to date. System manufacturers were allowed to develop proprietary features within the P25 standard. This has resulted in an "open architecture" that isn't as open as intended. This is especially true in large, complex multizone trunked systems, where complex proprietary roaming schemes are used to allow radios to operate over large geographic areas.

P25 Characteristics in High-Noise Environments

When P25 is used in settings where the background noise level is within limits set in the P25 standard, it provides useable audio. However the P25 vocoder was not designed to operate in the high-background-noise environments encountered on the fireground. When the P25 vocoder was being developed, the designers tested intelligibility of the digital audio with high ambient noise levels at the receiving radio. The P25 vocoder is unable to differentiate the spoken voice from the high background noise and assigns a digital value that does not accurately represent the voice. The result is unintelligible audio or broken audio with digitized noise artifact. Users of P25 radios have been affected by many common fireground noises. The SCBA and alerting systems for low-air or inactivity and PASS (Personal Alert Safety System) devices have made the audio transmitted from digital radios unusable. P25 radios transmitting from high-noise environments do not perform to the same levels as analog radios.

Self-Contained Breathing Apparatus Mask Effect on Communications

The effect of SCBA masks on the human voice has been studied by the Institute of Electrical and Electronics Engineers (IEEE).[2]

The IEEE tests were performed to find the effects of the SCBA system on voice intelligibility. Tests were performed with no SCBA mask and with an SCBA mask, with analog audio and digital audio. Participants wearing the SCBA masks read standard word recognition sentences while the listeners recorded what they thought they heard. As you will note in the table below, digital word error rates (WER) were always higher than analog error rates. Digital in Mask A had an average WER of 12.5 percent and digital in Mask B had an average WER of 6.8 percent. All of these tests were performed in a sound studio with no background noise. In an actual firefighting situation, the WER likely would be higher. Tests were also performed with the SCBA low-air alarm from each manufacturer in operation. In these tests the WER averaged 18.7 percent in analog and increased to 64.4 percent in digital.

Speaker	No mask Analog WER%	No mask Digital WER%	Mask A Analog WER%	Mask A Digital WER%	Mask B Analog WER%	Mask B Digital WER%
Male 1	1.3	1.8	5.3	12.0	5.5	6.1
Female 1	2.7	3.5	4.8	11.4	3.7	6.9
Male 2	1.6	3.1	4.2	14.2	4.2	7.3
Average	1.9	2.8	4.8	12.5	4.5	6.8

P25 Digital for Firefighting?

Fire departments around the country have reported difficulties with digital radios, and studies performed by National Institutes of Standards and Testing (NIST), the International Association of Fire Chiefs (IAFC), and portable radio manufacturers have supported the findings from the field users. Based on the experiences of fire departments using digital radios and the studies in response to these problems, the International Association of Fire Fighters (IAFF) has taken a position that does not recommend P25 digital portable radios for fire-fighting applications where the firefighter is using an SCBA.

These studies cite decreases in the ability of the firefighter's voice to be translated into a digital signal by the P25 radios. When fireground noise is introduced, the voice translation ability of the P25 radio provides decreased to no intelligibility. These problems are worsened when the firefighter is speaking into the portable radio through an SCBA facepiece, with or without a microphone inside the facepiece. Bone microphones or throat microphones may minimize the interference caused by background noise but are impractical for most firefighting portable radio uses. Speaker microphones are subject to the same problems that are found with the microphone on the portable radio.

The configuration of the P25 vocoder is limited in its capability to translate the human voice in the presence of common fireground noise or through a facepiece. This can pose a safety hazard for fireground operations. To maintain safety, fire departments should consider using portable radios that incorporate analog modulation for operations where the firefighter is using an SCBA.

Radios using the P25 digital technology have performed well for other fire service functions, such as on emergency medical incidents, support functions on the fireground where an SCBA is not required, and law enforcement operations. The difficulties presented by the inability of P25 radios to produce intelligible voice messages in the presence of fireground noise is a significant safety concern and should be considered seriously by public safety radio system designers and users.

The communications industry is aware of the present operational problems with P25 vocoders. If this issue is addressed and corrected by the industry, future P25 radios may be suitable for firefighting operations. Another area of opportunity for improvement is for the portable radio function to be more completely integrated into the SCBA. This integration may lessen the impact of background noise on the vocoder.

Direct and Repeated Radio Systems

Radios communicate when the transmitter sends out a signal that is received by one or more receiving radios. When the signal is received from the radio initially transmitting the signal, the communication is direct (i.e., there is no intervening radio or system). One radio transmits, the other radios receive, and this type of communications also is known as simplex communication.

Nonsupported Simplex Communications on the Fireground

Using simplex communications maintains positive communications between the IC, exterior onscene units, and interior units without the reliance on exterior communications systems. Maintaining positive communications is especially important in "Mayday" situations. When users on simplex radios are deployed to the interior of a structure they create a radio receiver network. As more and more radios move into the structure, the strength of the network increases. If Engine 1 calls Mayday, the probability of another radio on the interior receiving the transmission is high. If the Mayday is not heard by the IC, another radio operator on the interior can act as a human repeater to repeat the message to the IC. In addition, the number of radios in a structure creates redundancy, where reliance on a single repeater or trunked system creates a single point of failure. Simplex communications allow direct communications with the initiator of the Mayday and other crews on the fireground.

In this example, the simplex communications are not supported. This means that there is no infrastructure to support transport of the fireground communications to the dispatch center. When the radios involved in direct communication are portable radios, the communication distance typically is limited to a few miles; for mobile radios the distance can be 50 to 100 miles. Often this is referred to as "line-of-sight communication" and this makes direct radio communication most suitable for use by units on an incident scene.

Figure 17 – Simplex Fireground Communications.

Non-Supported Simplex Fireground Communications

Pump 1 Receives

Dispatch Center Out of Receive Range

E 4 Receives

E 1 Transmits

E 2 Receives

Pump 3 Receives

E 3 Receives

The cloud represents the maximum transmit range of the portable radio

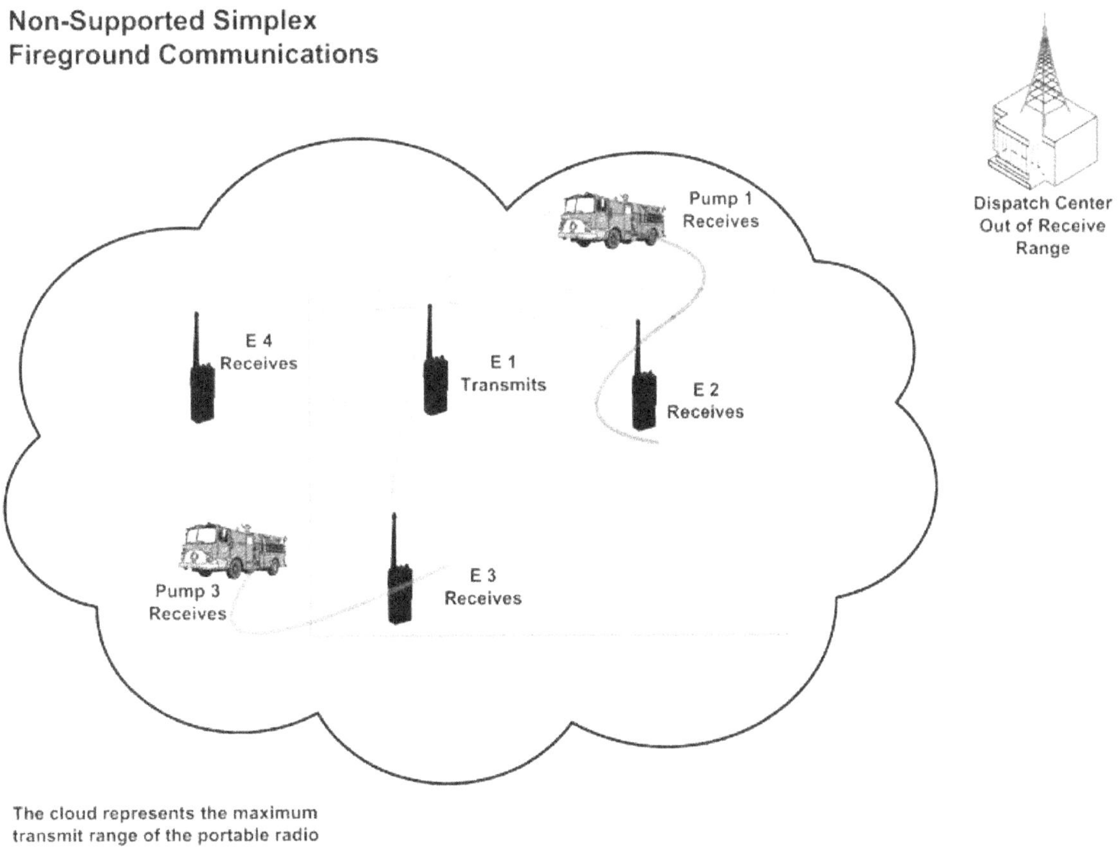

The direct communication method is the simplest form of radio communication and is easily affected by terrain blocking. If a mountain or other obstruction is between the transmitting and receiving radios, communication may not be possible. However, the short-range nature of direct communication also allows the radio channel used by one communicating group to be reused by another group further away. If the second group is far enough away that it does not hear the first group's communications, then the channel can be reused. This minimizes the number of channels needed by an agency.

When a radio system must cover a larger area, or when terrain or other obstructions limit the distance a system can cover, additional equipment is needed to overcome these limitations.

Receiver Voters — Improve Field Unit to Dispatcher Communications

Dispatch centers connected to high-powered transmitters provide the dispatch center with talk-out capability. Transmitters are elevated to achieve better line-of-sight communications with the service area. High-powered transmitters ensure that the dispatch center transmissions are heard throughout the service area and provide some level of in-building coverage. See **Figure 18**.

Portable radios have limited power and cannot always transmit a signal strong enough to reach the transmitter sites. To provide a more balanced system, receivers are networked together throughout the service area in a receiver voter system (RVS). Comparison of the received audio signal takes place in a **receiver voter.** The receiver voter and its network of receivers are referred to as the **RVS**. The RVS usually is located at the dispatch center. The receiver voter compares the audio from all receivers and routes the audio from the receiver with the best audio quality to the dispatcher. This type of system provides very reliable fireground communications and supports fireground simplex channels.

Figure 18 – Simplex Fireground Communications with Dispatch Center.

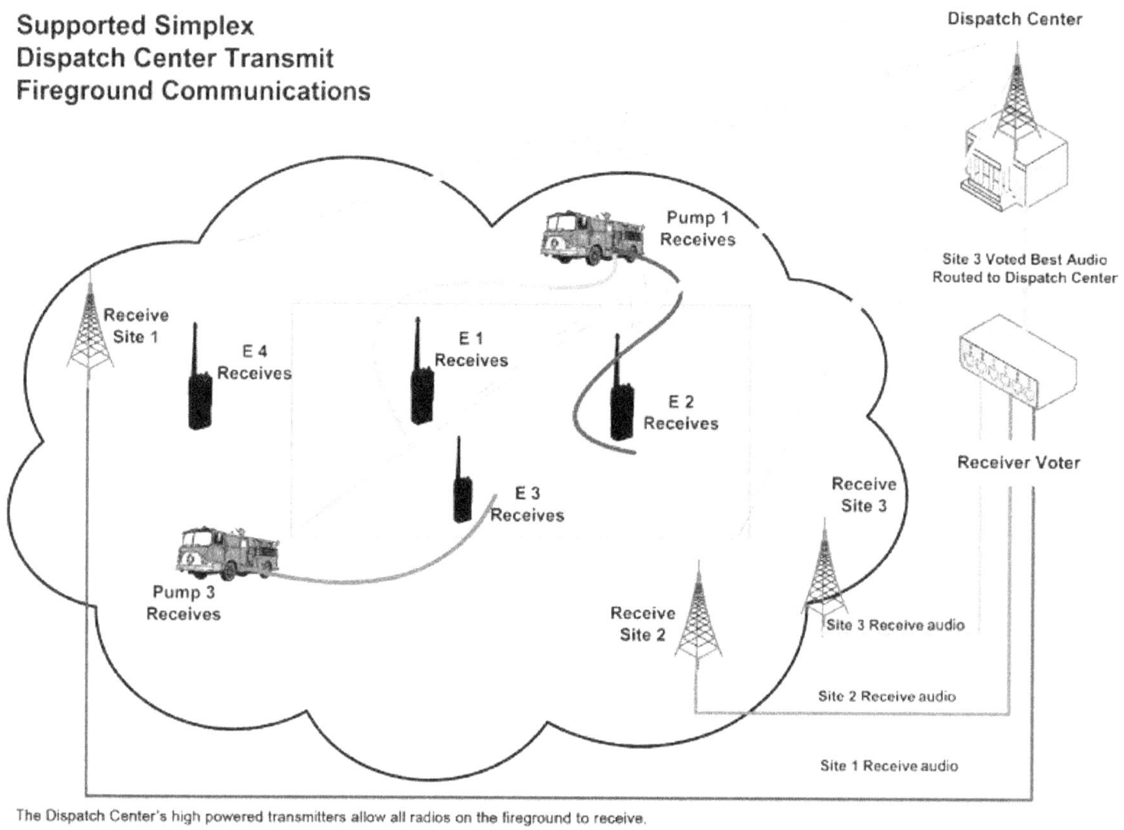

The Dispatch Center's high powered transmitters allow all radios on the fireground to receive.

Figure 19 – Simplex Communications with Portable Transmitting.

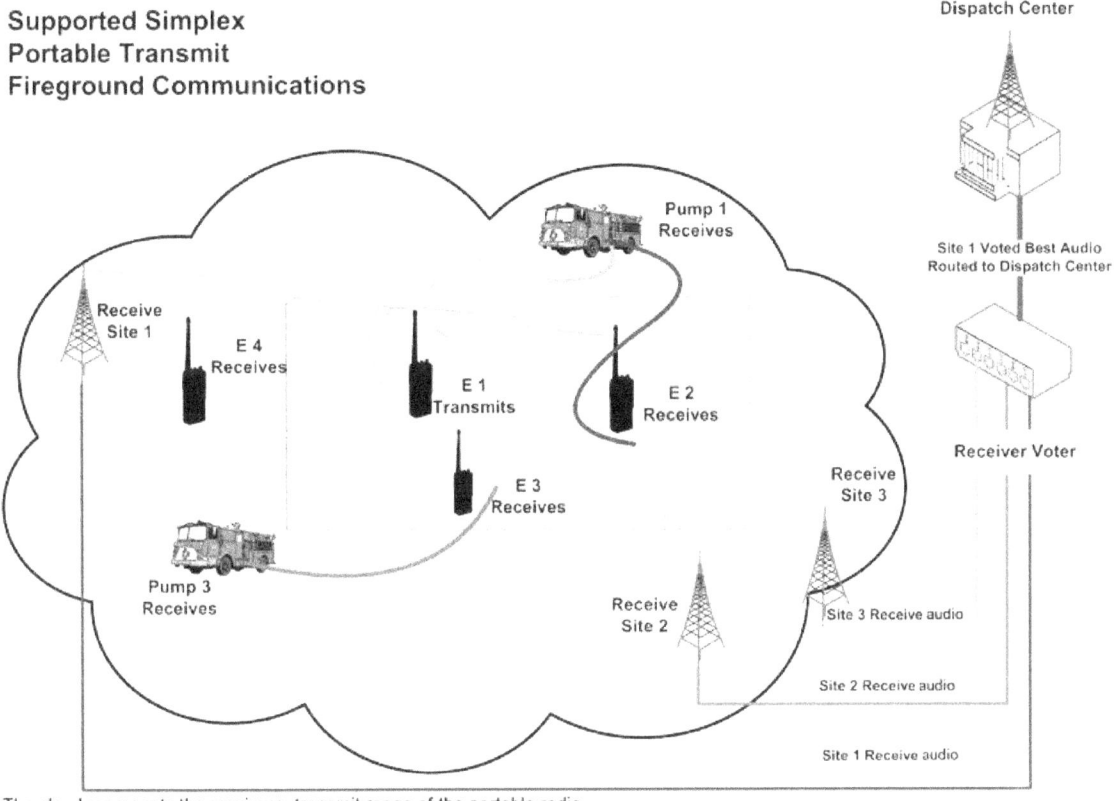

The cloud represents the maximum transmit range of the portable radio.

Repeaters — Improve Field Unit to Dispatch and Offscene Units

Receiver voters are one solution to get communications from a radio user to the dispatch center, but another solution is needed to get the communication to other radio users. One type of system that can solve this problem is a repeated radio system. Repeated radio communication, also known as half duplex communication, uses two radio frequencies for communication. The transmitting radio transmits on frequency 1 (F1), and that signal is received by the repeater. The repeater then repeats the transmission on frequency 2 (F2), and this signal is received by the receiving radio. By locating the repeater on a high building or mountain, the range of transmissions from the transmitting radio can be more than doubled, and can reach over obstacles effectively.

Another solution to improving communication between field units inside buildings or tunnels and dispatch and offscene units is the bidirectional amplifier (BDA). BDAs can be used with half duplex radio systems to extend coverage from inside the structure to the outside of the structure and vice-versa, but BDAs do not operate with simplex radio systems. BDAs are discussed in more detail in Section 5—Trunked Radio Systems.

Figure 20 – Half Duplex Fireground Communications.

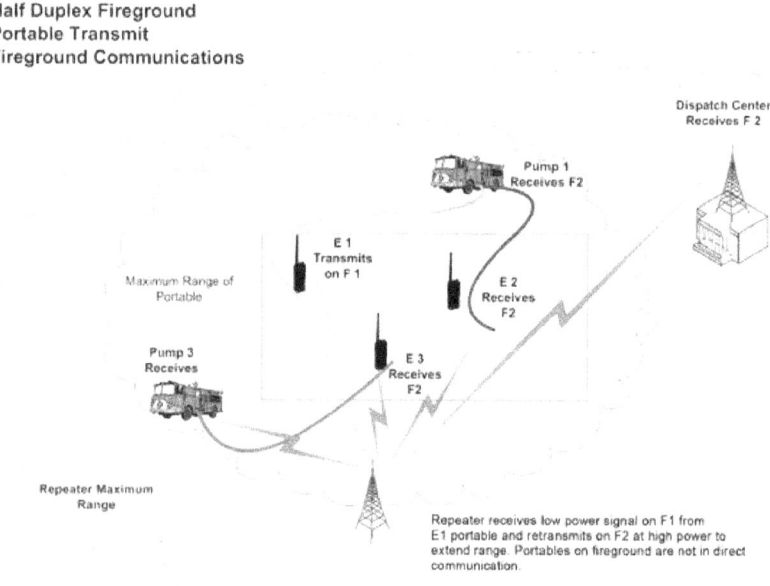

The significant operational difference between direct and repeated communications systems affects units operating at an incident scene. With direct communication, the transmitting radio's signal only needs to reach other radios directly on the incident scene. With a repeated system, the signal must reach the closest repeater location, which may be much further from the incident than the receiving radios.

Figure 21 on the next page, shows a method to overcome this limitation. If unit E1 is unable to communicate with other units on the fireground using the repeater system, E1 can switch to talkaround mode on the radio. This mode allows the unit to transmit in direct mode to other radios on the fireground and receive from the units in either direct or repeated mode. Since the radio is not able to reach the repeater, the dispatch center cannot hear the radio, although other radios on the fireground can hear the unit. A unit that switches to talk-around should announce this immediately so other units know that they also may need to switch to communicate with the isolated unit.

Figure 21 – Half Duplex Communications with Talkaround.

Simulcast Transmitter Systems

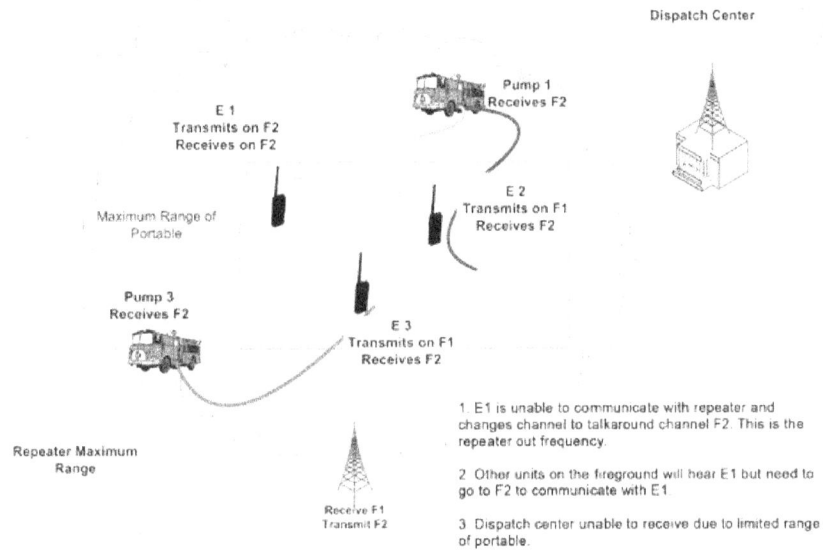

Half Duplex Fireground
Portable Transmit Talkaround
Fireground Communications

When a radio system must cover a large area, but the number of available frequencies is limited, a simulcast transmitter system may be the solution. With this system, multiple transmitters simultaneously transmit on the same frequency. The transmitters must be precisely synchronized so that the signals they transmit do not interfere with each other. In addition, the audio source sent to the transmitters must be synchronized so that the radio user hears the same signal from each transmitter. The system consists of a simulcast controller and two or more simulcast transmitters. The advantages of a simulcast system are the coverage of a large area, with high signal levels throughout the area, while using only a single frequency.

Figure 22 – Simulcast Transmission from Dispatch Center.

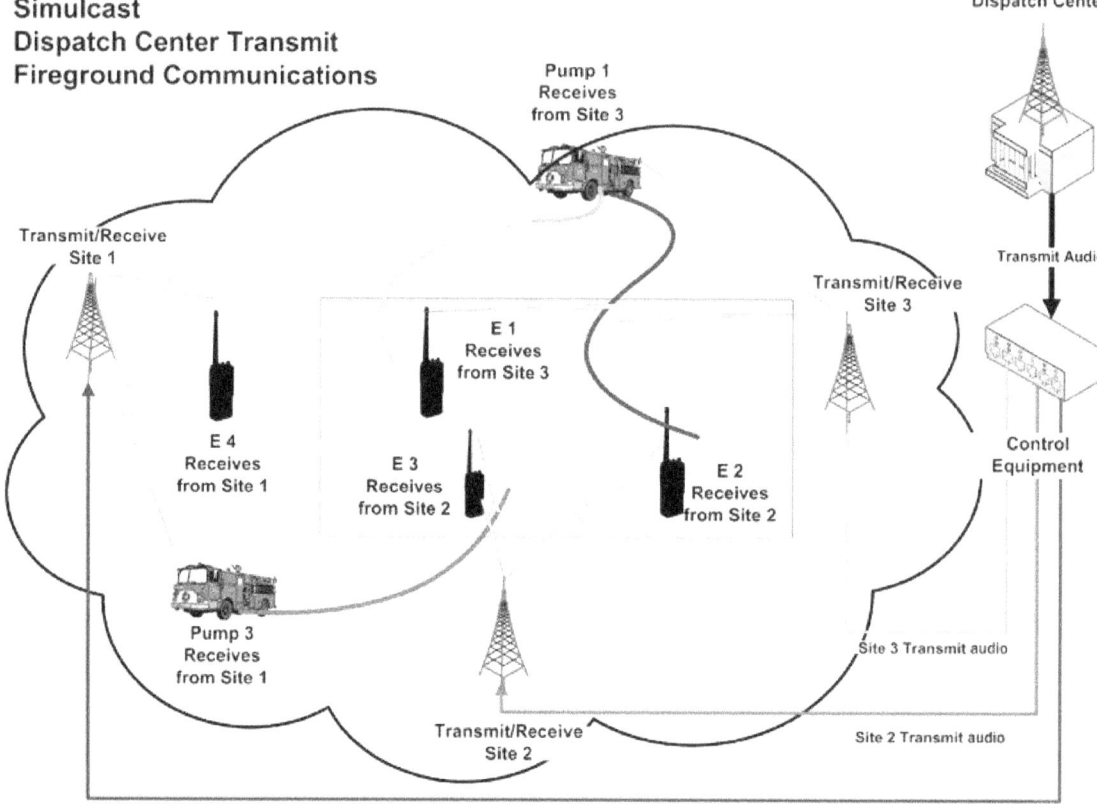

Operational Considerations

Communication needs on the fireground can be categorized based on the position in the Command structure. The military operates in a similar manner and is used here to illustrate the concept.

In a military theatre of operations, there are distinct communication requirements based on position in the Command structure. At the lowest level are individual team communications. This level of communication is for command and control of the team members to accomplish a task. Like the fire service, these communications are often simplex communications and short range. An example of this would be communications over a SCBA intercom—a short-range, low-power radio designed for communications among a single company or crew.

Figure 23 – SCBA Intercom.

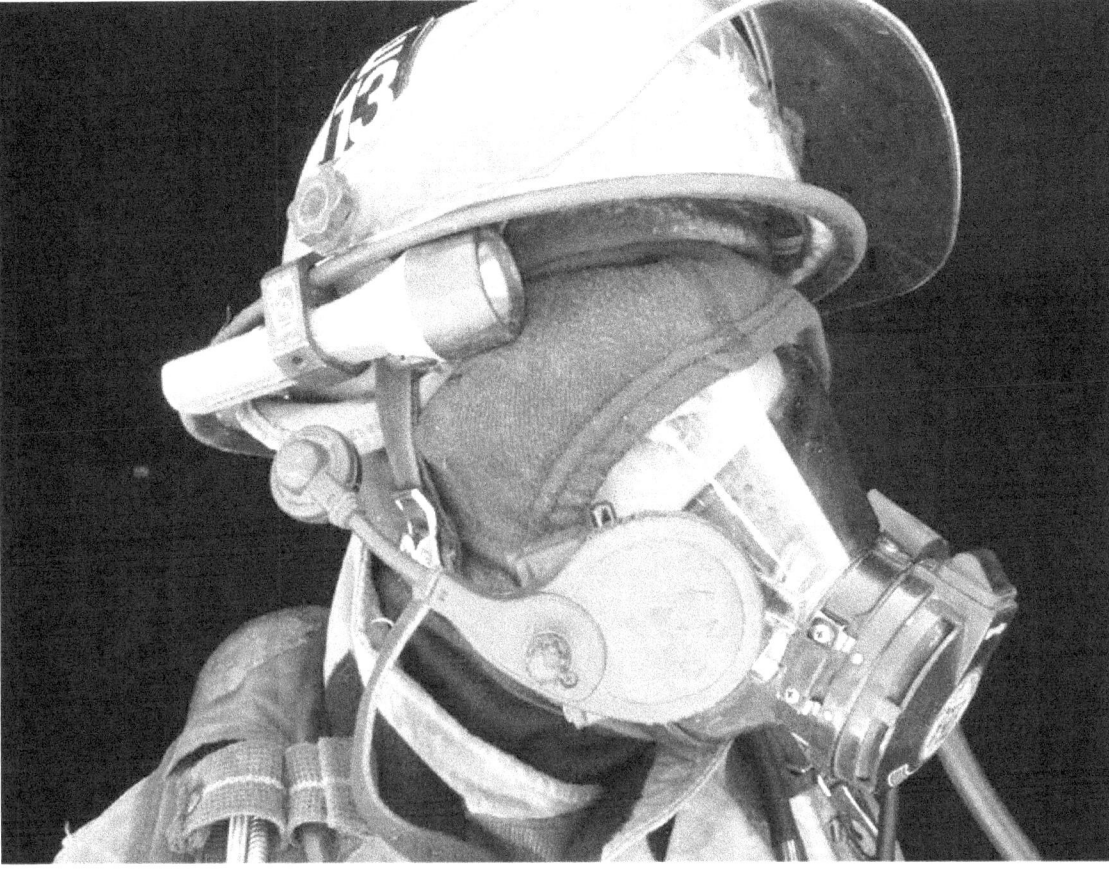

Next is the tactical level, where the tactical Command element is communicating with multiple teams to accomplish the tactical objective. This level requires communications with enough range to communicate with all teams assigned to achieve the tactical objective. The equivalent in the fire service would be the fireground tactical radio channel or talkgroup where the IC is coordinating a fire attack with several companies.

The last level is the strategic level, where a military commander is responsible for several areas of operation. This requires wide area communications to communicate with each tactical commander in his/her area of operation. In the fire service, the strategic-level communications commonly are wide-area communications to the dispatch center for requests for additional resources and documentation of tactical benchmarks. The dispatch center has the strategic responsibility of maintaining response capabilities in the unaffected areas of the city.

Additional communications layers should be considered based on the amount of radio traffic occurring on a radio channel and the complexity of the operation, if radio channels or talkgroups are available. When radio channel traffic increases to the point that the channel becomes saturated, this, in itself, becomes hazardous. If a Mayday or other emergency occurs there is no reserve capacity on the radio channel to handle the event. Complex operations often require immediate radio communications free of other radio "chatter," and should be assigned a separate channel.

Fireground communication systems should address the operational levels where communications are occurring. A common practice is to assign responding units two radio channels when dispatched. One channel is designated a Command channel, and the second is the tactical channel. The Command channel provides the IC with a wide area channel to communicate with the dispatch center. The tactical channel is a simplex channel for fireground communications with crews assigned to interior fire attack. This is a pure separation of the tactical and strategic levels. Other departments often mix the strategic and tactical levels by having the dispatch center monitor and document tactical-level communications on a single channel.

SECTION 4
PORTABLE RADIO SELECTION AND USE

General

The success of a fire service radio system project hinges on the performance of the portable radio. If the portable radio has poor performance, the end-user relates it to the performance of the radio system as a whole. All the firefighter knows is that when the PTT was pressed the communications worked or did not work.

Manufacturers offer radios at different price points to meet market need. As with any other product, the options and performance levels increase with the cost. Usually there are three tiers of radios available. At the lowest level are nonruggedized radios meant for users who do not handle radios in a rough manner and do not operate in environmental extremes. The second level of radio is for the user who needs more reliability and performance features. The highest tier radios are focused on the public safety user. They offer the highest levels of performance and reliability and have the most options available. At this level, the radios often are submersible and have intrinsically safe options. Submersible radios are a very worthwhile option for the fire service, considering the possibility of radios getting wet or exposed to steam.

Ergonomics

Today's radios are an integral part of firefighting and a key component of fireground safety. The form and fit of the radios for firefighting has not improved much over the past decade. Buttons and knobs have increased in size as compared to the radios of the 80s and 90s, but firefighters have the same difficulties operating radios while in personal protective equipment (PPE). Radio knobs still cannot be manipulated with a gloved hand, even though it is required as a component of National Fire Protection Association (NFPA) Standard 1221, *Standard for the Installation, Maintenance, and Use of Emergency Services Communications Systems.*

> **9.3.6.6** Portable radios shall be designed to allow channels to be changed while emergency response personnel are wearing gloves.

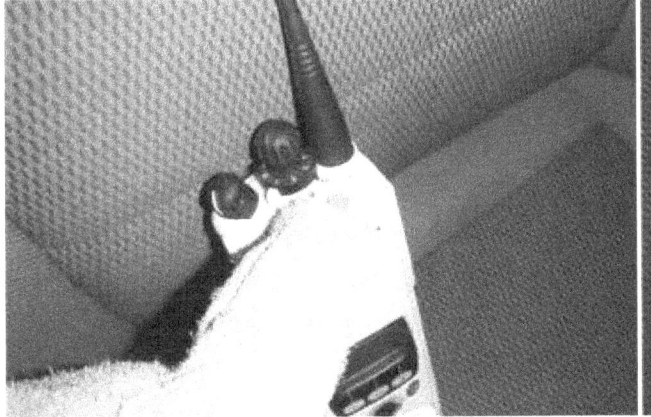

The radios of today can be programmed with hundreds of channels or talkgroups. The large number of channels/talkgroups has made "hard switches" that correspond with a channel/talkgroup impossible. To select channels on radios with added channel capabilities requires liquid crystal displays (LCD) and "soft keys" to provide access. In firefighting, the LCDs are not readable in smoky environments and the soft keys cannot be pressed with a gloved hand. When programming the radio, take care to make firefighting radio channels easily accessible.

Environmental Technical Standards

Radios are designed to operate in environmental ranges. The harsh environment of firefighting is hard on equipment and personnel. To provide reliable communications, it is common to purchase ruggedized communications equipment. The technical specifications and testing protocols used to determine if a device is rugged can be confusing. Manufacturers use several testing protocols to determine if the device is "Public Safety Grade." Some of the more common standards encountered are Military Standards (Mil Std) and International Electrotechnical Committee (IEC) standards.

IEC IP (Ingress Protection) Codes

IP codes are international standards that test for ingress protection into an electrical enclosure. Manufacturers use this code to rate intrusion against solid objects from hands to dust and water in electrical enclosures. The rating consists of the letters IP followed by two digits. The standard is intended to provide an objective testing protocol to reduce subjective statements such as "waterproof". The first digit represents the size of the object that is protected against and the second digit represents the water protection. More detailed information on this standard can be found at www.iec.ch, International Electrotechnical Committee, IEC 60529.

Mil Standards

In the 70's and 80's radios were manufactured to various industry standards for ruggedness and technical stability. In the 90's radio manufacturers adopted Mil Std 810 as a standard for reliability and ruggedness. Mil Std 810 was developed by the military to provide an environmental test protocol that would prove qualified equipment would survive in the field. Mil Std 810 is a test protocol written for the military environment not the firefighting environment. The specification sheets often reference a letter designation behind the Mil Std. The letter designation represents the revision level of the Mil Std being tested to. The latest revision is Mil Std 810 F. Earlier revisions of the Mil Std 810 were generic up to revision C. Subsequent revisions became more tailored to the actual environment the equipment would operate in. Manufacturers sometimes only perform specific test components of the Mil Std. For instance, an equipment specification may read "Mil Std 810 F for water, dust and shock resistance". When we see Mil Std 810 we assume that the equipment is ruggedized and will survive the firefighting environment. We need only look to the temperature specification to see that this is questionable. Mil Std 810 F actually has two temperature specifications depending on where the equipment is to be used.

Mil Std 810 F High Temperature Table.

Design Type	Location	Ambient Air °C (°F)	Induced [2] °C (°F)
Basic Hot	Many parts of the world, extending outward from hot category of the United States, Mexico, Africa, Asia, and Australia, southern Africa, South America, southern Spain, and southwest Asia.	30 - 43 (86 - 110)	30 - 63 (86 - 145)
Hot	Northern Africa, Middle East, Pakistan and India, southwestern United States and northern Mexico.	32 - 49 (90 - 120)	33 - 71 (91 - 160)

The table shown is the high temperature table from Mil Std 810F. A similar table is included in Mil Std 810 F for low temperatures. Most manufacturers test to the "Basic Hot" and "Basic Low" temperature levels. This temperature range is from approximately -30° C to 60° C (-22° F to 140° F). These temperature extremes do not replicate the environments that firefighters encounter.

National Institute of Standards and Technology Testing

NIST has performed testing on portable radios that more closely mimics the firefighting environment.[4]

> The results of these tests exposed the vulnerability of the portable radios to elevated temperature conditions, and emphasized the need to protect the radios when used in firefighting situations. Radios tested inside the turnout gear pocket showed that the turnout gear pocket was able to protect the radios and allow them to operate at the Thermal Class III temperature of 260 °C. This contrasts with tests where the radios were exposed directly to the airflow, in which the radios did not survive at Thermal Class II conditions and beyond. In all but one test, the exposed radios were able to operate properly at the Thermal Class I temperature of 100 °C, above the listed maximum operating temperature of 60 °C. Failure of the electronics due to heating was not permanent for the radios. In all cases where the radio casing was not damaged, the radios regained normal operating function once they had sufficiently cooled. Permanent damage to the casing, such as difficulty turning knobs or pressing buttons did occur for some radios whose casings experienced melting. Permanent damage also occurred to the external speaker/microphones, especially due to the melting of the connecting cables.

> The next step for this project is to work with the NFPA to develop a radio standard that would include requirements for the thermal testing of handheld radios.

How Many?

After defining the technical and operational requirements of the radio, the number of radios needed has to be determined. Departments have to identify who needs radios. A portable radio for each firefighter provides the highest level of safety. In addition to firefighters, radios for support and other fire department functions should be considered.

Additional guidance can be found in the following NFPA standards:

• NFPA 1561, *Standard on Emergency Services Incident Management System*:

    - **6.3 Emergency Traffic**.

        - **6.3.1*** To enable responders to be notified of an emergency condition or situation when they are assigned to an area designated as immediately dangerous to life or health (IDLH), at least one responder on each crew or company shall be equipped with a portable radio and each responder on the crew or company shall be equipped with either a portable radio or another means of electronic communication.

- NFPA 1221:

  - **9.3.6 Two-Way Portable Equipment.**

  - **9.3.6.1** All ERUs (Emergency Response Units) shall be equipped with a portable radio that is capable of two-way communication with the communications center.

What Type?

Since radios are tiered based on performance and ruggedness, there can be significant cost savings by buying high-tier radios for responders and the appropriate lower tiered radios for support staff.

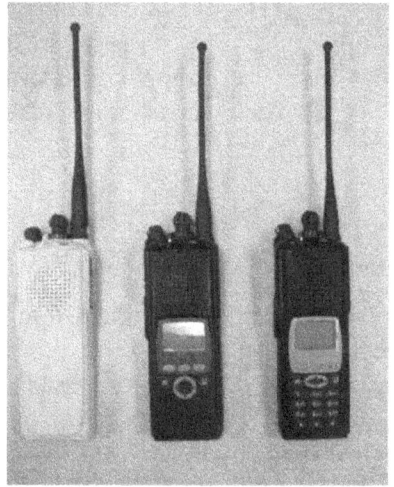 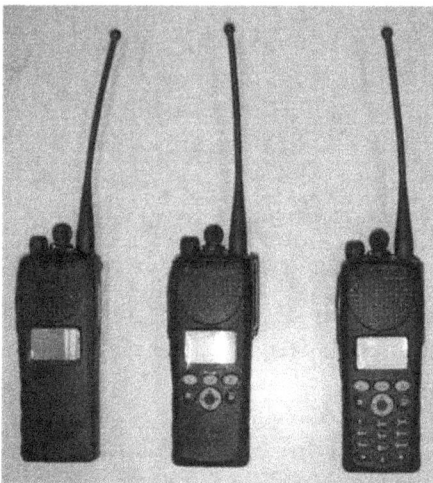 

High-tier Radios   **Midtier Radios**   **Low-tier Radios**

- High-tier — High-tier radios should be provided to each firefighter. This level of radio gives the highest level of performance and reliability that radio manufacturers can provide. Within each tier there may be options that provide additional capabilities or functions. If using radios for emergency medical services (EMS) and fire functions, encryption may be required for operations with law enforcement agencies or to comply with the Health Insurance Portability and Accountability Act of 1996 (HIPAA) requirements.

- Midtier — Midtier radios may be appropriate for users who do not enter into the firefighting environment. This type of radio would be a good choice for EMS functions. Again encryption may be required to meet HIPAA requirements.

- Low-tier — Low-tier radios are an option for some support staff. These radios provide communications for users who are not in harsh environments and may not need all the functionality of the higher tiered radios.

Fire Radio Features

Many features are available in modern radios. Like automobiles, stripped-down versions of radios are available, but when options are added the cost rises. To identify the desired features, focus and user groups can assist in developing the radio feature sets that meet user's needs. Today's radios are extremely flexible in programming features and the functions of buttons on the radio. Cooperation between the radio vendor and technical provider for your radio system will be instrumental in filtering through all of the programming parameters. Some of the newer features that increase firefighter safety are

- Voice channel announcement — This feature uses prerecorded voice prompts to notify the firefighter what channel the radio is on as the channel select knob is moved.

- Emergency indications — Radios on the fireground receive an indication of emergency activations on the assigned channel.

- Personnel accountability — In new systems there are more radio ID numbers available. This makes it possible for each radio to have an individual ID code enabling identification of the unit and specific position of the unit on an emergency activation. If tied to roster information in a computer-aided dispatch (CAD) system, identification of the individual firefighter is possible.

- Tones — Many radios use tones as an indication of trunked system access, out of range, repeater access, encrypted channel, and others. Use of tones may provide added awareness to the firefighter and, thus, increase safety.

For guidance on the minimum feature set a radio should have, refer to NFPA 1221 Section 9.3.6:

9.3.6 Two-Way Portable Equipment.

9.3.6.2 Portable radios shall be manufactured for the environment in which they are to be used and shall be of a size and construction that allow their operation with the use of one hand.

9.3.6.3 Portable radios equipped with key pads that control radio functions shall have a means for the user to disable the keypad to prevent inadvertent use.

9.3.6.4 All portable radios shall be equipped with a carrier control timer that disables the transmitter after a predetermined time that is determined by the authority having jurisdiction.

9.3.6.5 Portable radios shall be capable of multiple-channel operation to enable on-scene simplex radio communications that are independent of dispatch channels.

9.3.6.6 Portable radios shall be designed to allow channels to be changed while emergency response personnel are wearing gloves.

9.3.6.7 Single-unit battery chargers for portable radios shall be capable of fully charging the radio battery while the radio is in the receiving mode.

9.3.6.8 Battery chargers for portable radios shall automatically revert to maintenance charge when the battery is fully charged.

9.3.6.9 Battery chargers shall be capable of charging batteries in a manner that is independent of and external to the portable radio.

9.3.6.10 Spare batteries shall be maintained in quantities that allow continuous operation as determined by the authority having jurisdiction.

9.3.6.11 A minimum of one spare portable radio shall be provided for each 10 units, or fraction thereof, in service.

Portable Radio User Guide

Users and their behaviors have an impact on the effectiveness of fireground communications. Human factors, such as the way we speak and organization of reports, affect communications. Technical factors obviously have an impact on fireground communications. Like any other technology, users need to know the limitations of the technology and how to use the tool appropriately.

Human Factors

When we talk on the radio, each of us subconsciously performs a process before we speak. Managing this process will provide more effective communications.

- Organization — Before speaking, formulate what information is being communicated and put the information in a standardized reporting template. For instance, a standard situational report might contain Unit ID, location, conditions, actions, and needs. This method forces users to fill in the blanks, answer all the necessary questions, and filter out unneeded information.

- Discipline — Often, ICs are overwhelmed by excess information on the radio. Radio discipline on the fireground will help to determine if information needs to be transmitted on the radio. If face-to-face communications are possible between members of a crew and the information is not needed by the IC, don't get on the radio.

- Microphone location — Placing a microphone too close to the mouth or exposing the microphone to other fireground noise may result in unintelligible communications. When transmitting in a high-noise environment, shield the microphone from the noise source. Hold the microphone a couple of inches from the mouth or, when speaking through a SCBA mask, place the microphone near the voice port on the facepiece.

- Voice level — When speaking into a microphone use a loud, clear, and controlled voice. When users are excited, the speech often is louder and faster. These transmissions often are unintelligible and require the IC to ask for a rebroadcast of the information, resulting in more radio traffic on the channel.

Managing these human factors will have a positive impact on fireground communications. Reporting should be complete, necessary, and in a controlled, clear voice. These actions will reduce the amount of repeat transmissions on the fireground, reducing air time.

Technical Factors

In some cases communications problems are caused by a technical issue. Users need to recognize technical problems and take corrective action to improve communications. Radio users often blame the radio or system for coverage problems. In many cases user actions can improve communications.

Position and Radio Location

As we know, firefighting often places firefighters in challenging environments forcing them to crawl on the floor. The optimal position for a portable radio transmission is at head height with the antenna in a vertical position. The photo below shows a firefighter on the floor. The radio is against the body inside of the radio pocket on the turnout coat. Some of the transmitted energy is absorbed by the body, and the antenna is in a horizontal position. This results in a poor radiation pattern and a reduction in range of the radio. Moving to a location where the firefighter can sit up may improve communications, if transmissions from the prone position are not heard.

Many users do not use a radio pocket or case. In the middle photo above, the Company Officer's (CO's) radio (left) is clipped to the exterior of the coat, while the firefighter's radio is protected. The tradeoff is that the radio is exposed to heat and steam but is in a better transmitting position. When unprotected, the radio may fail to operate when needed. Radio cases with shoulder straps provide little protection and are an entanglement hazard when worn on the exterior of turnouts.

Coverage

When communicating on the fireground some areas of a building may be difficult to communicate from. When encountering these areas, move to a location where communications are possible. Areas that may improve communications are near windows and doors.

Accessories

Many accessories are available for radios. Use of accessories that protect the radio from heat and steam allows the radio to operate in high heat environments. Each user group may have specific working conditions that require some accessories to make radio communications easier. Common accessories include carrying cases, speaker microphones, ear pieces, chargers, battery types, and optional antennas. User and focus groups will help identify the accessories needed to support your department's needs.

Summary

Many factors contribute to the success or failure of fireground communications. Some are human factors that affect the way information is processed and communicated. Processing and communicating information in a standardized manner assists in managing information and the amount of traffic on the radio. Speaking with a loud, clear, and controlled voice will reduce the amount of repeat radio traffic on the fireground. Technical factors require recognition of the problem and some corrective action by the user to improve communications.

The portable radio equipment is what the firefighter sees and has the most impact on the perceived performance of a system. Inclusion of user groups in the selection process of the portable radios and accessories provides insight into the functions and features needed. Determine the size of the radio fleet and the users who need radios. This will assist in classifying each user type and the radio tier needed for each user classification. Selecting the correct tier radio for the target user is a way to contain cost and provide reliability for users on the fireground. While today's radios provide reliable communications, they are not manufactured for the extreme environments of the fire service. When selecting radios, consider features that increase firefighter safety. The minimum feature set for firefighting portables can be found in NFPA 1221.[5]

SECTION 5
TRUNKED RADIO SYSTEMS

Trunked radio systems are complex radio systems that were developed to improve the efficiency of the use of available radio spectrum. In conventional (nontrunked) radio systems, a radio frequency is dedicated to a single function or workgroup. When the radio frequency is not in use, it cannot be used by another function or workgroup. Trunking borrows technologic concepts from telephone systems to assign radio frequencies to active calls, improving the efficiency of frequency use.

Like a conventional repeated radio system, trunked radios communicate with each other through two or more repeaters. In a trunked system, the radios often are known as **subscriber units** and a voice communications exchange is know as a **call**. A basic trunked radio system has a system controller that controls the assignment of the repeaters, called voice traffic repeaters, to individual calls. The radios communicate with the system controller, for example to request the use of a voice traffic repeater, by sending data messages to the system controller on a special dedicated channel called the control channel. The system controller acknowledges these communications and sends information to the radios using the control channel as well. The radios also can communicate some information using the voice traffic channels after a call has been terminated.

Figure 24 – Trunked Radio System.

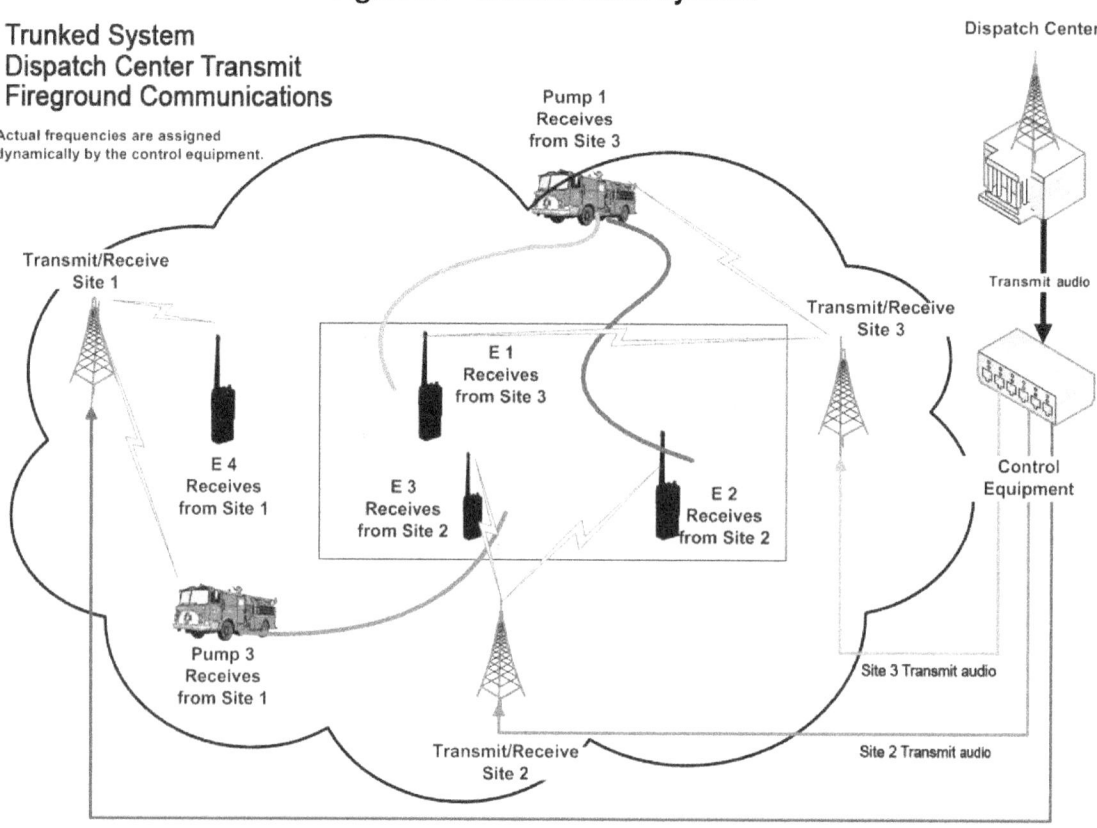

The voice traffic repeaters are shared among all users of the system; they also are known as **resources**. In complex systems that use encryption and dispatch consoles, other equipment is necessary for the operation of these features, and they are considered **shared resources**.

The radio industry uses the term **talkgroup** to distinguish among physical frequencies or channels used in conventional radio systems. This terminology often is confusing, since from the actual radio user's point of view a talkgroup and a conventional channel are the same; they are both communications paths. The distinction is made by the technologists to differentiate a physical channel or frequency from the logical channel or talkgroup.

The system controller and other parts of the trunked radio system maintain a log of all activity that occurs in the system, as well as statistical information on the operation of the system. These system logs can be used in the event of a suspected anomaly in the operation of the system to help determine the cause.

General Radio Operation

Radio On/Off — Registration/Deregistration/Talkgroup Affiliation

When a trunked radio is powered on initially, it begins operation by telling the system controller that it is active, along with the talkgroup currently selected on the radio, using the control channel. If the registration is successful, the radio is registered on the system and now can receive and transmit; if the registration is not successful, the radio will not operate on the system.

Any time that the radio is powered on and the user changes talkgroups the radio will tell the system the new talkgroup selection, and the system will confirm the selection. In this way, the system tracks the currently selected talkgroup for all radios registered on the system.

When the radio is switched off by the user, the radio transmits a message to the system controller telling the system to deregister the radio. The radio then will wait for an acknowledgment from the system before actually powering off.

Talkgroup Call

When a radio user wishes to transmit on a talkgroup, he/she presses the PTT switch, just as with a conventional radio. The radio then sends the trunking system a request to transmit, using the control channel. The trunking system checks to see if the requested talkgroup is free and if there are available voice traffic repeaters. If these are true, then the system assigns a voice traffic repeater to the call and instructs all radios with the talkgroup selected to change frequencies to the voice traffic repeater frequency. The system also sends a message to the requesting radio telling it that it may proceed with its transmission. This causes the user's radio to play a tone sequence (typically three short beeps) to tell the radio user that he/she may proceed with the transmission. The radio's transmission is received by the voice traffic repeater and retransmitted to the other radios on the frequency.

Figure 25 – Trunked Fireground Communications.

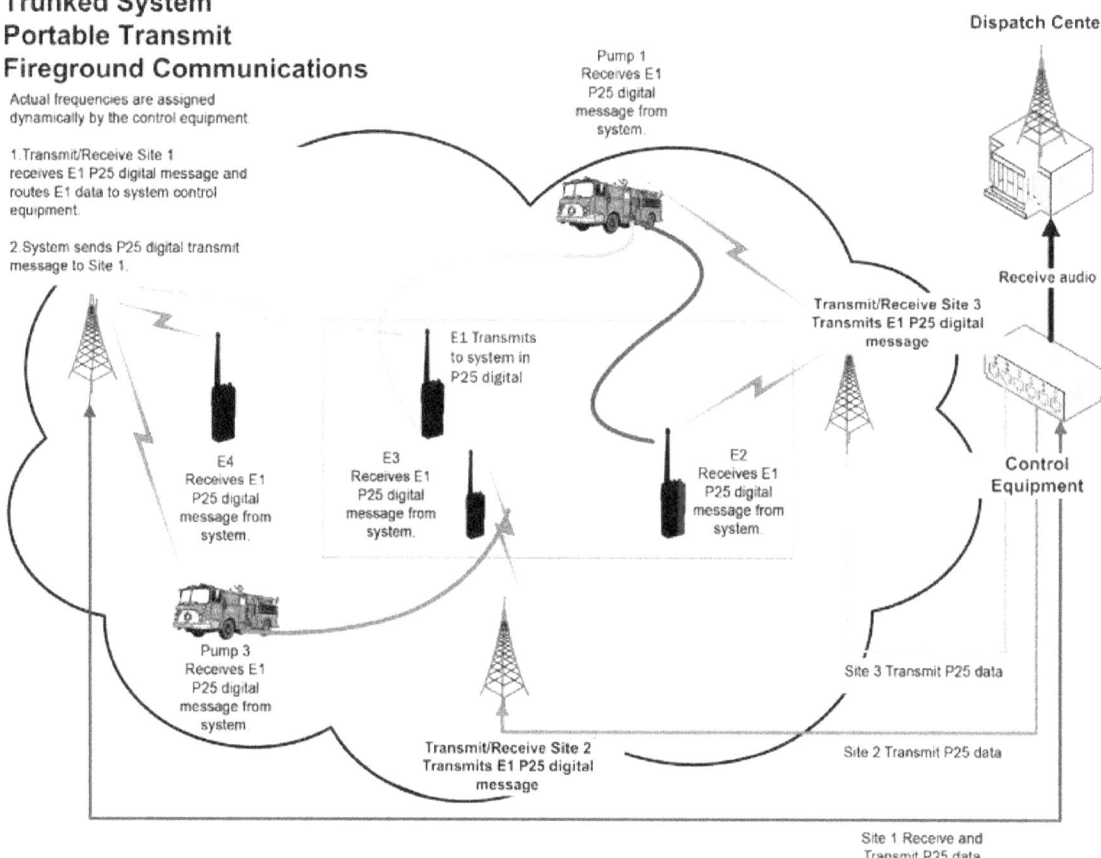

If the there are no voice traffic repeaters available for the call, the system will place the request in a busy queue in order of priority, send a busy message to the requesting radio and wait a short time for resources to become available. If the resources become available, the transmission proceeds. If the resources do not become available before the wait time expires, the system transmits a message to the requesting radio telling it that the request failed. The radio will play a tone (commonly called a "bonk") to the user indicating the failure.

Call Disconnection

When the transmitting user is finished with the transmission he/she will release the PTT switch. This causes the radio to send a message on the control channel telling the system that it can release the resources assigned to the transmission. Depending on the configuration of the talkgroup, the system either waits a few seconds for additional transmission requests before releasing the resources, or it releases the resources immediately. Once the timeout is reached, the system tells all radios on the talkgroup to change channels to the control channel and releases the voice traffic repeater for use for other requests. If another request is received before the resources are released, then the system immediately grants the requesting radio's transmission request and does not need to tell the other radios to switch frequencies.

Other Trunking System Features

Multigroup Call

A multigroup call is a call that transmits to two or more talkgroups simultaneously. The system can be configured to wait for all talkgroups in the multigroup to become available before initiating the call, or configured to begin the call immediately, with busy talkgroups joining when their calls are complete. During the call, all associated talkgroups act as a single talkgroup. Because of this, after the initial multigroup transmission completes, a user in one of the associated talkgroups can call all users in the associated talkgroups. In a busy system, this can keep the multigroup call in progress for a significant amount of time, severely disrupting operational communications.

Private Call

The **private call** feature allows one radio to call another radio and to carry on a conversation without any other radios hearing the conversation. The radio user initiating the call must select the called radio from a list, or know the numerical ID of the called radio. Some more advanced radios allow the user to change numbers in a cell-phone-like phone book, making this feature more usable.

A problem with the private call is that it is very difficult to predict the capacity or loading impact of this feature during system design. When the system is in operation, high private call usage can cause other system users to experience more talkgroup busy signals than the design would predict. Some system operators prohibit the use of private call to eliminate the possibility of these calls affecting more critical operations.

Emergency Alarm

There are two different emergency features in trunked radio systems: **emergency alarm** and **emergency call.** When a radio user presses the emergency button on the radio, the radio switches to the control channel and transmits an emergency alarm message. This message is processed by the system, and an indication of the activation of the alarm is presented to any dispatchers using radio consoles. The benefit of the emergency alarm feature is that it is possible to send the alarm message even when all repeaters in the system are busy. Thus, even when the talkgroup is in use, an emergency alarm can be sent by a firefighter in trouble.

Emergency Call

An **emergency call** is similar to a normal talkgroup call or a multigroup call, but the radio initiating the call is in emergency mode after having its emergency button pressed.

Emergency calls are initially processed in the same way as talkgroup calls or multigroup calls. The difference in processing occurs when resources are not immediately available for assignment to the emergency call. If resources are not available, the emergency call can be processed in two ways: top-of-queue or ruthless preemption, depending on the configuration of the trunked radio system.

If the system is configured for top-of-queue, the request for resources is placed on the busy queue in front of all other requests. When the resources become available, the emergency call is assigned the newly-available resources immediately.

If the system is programmed for ruthless preemption, the request for resources is not queued and instead the voice repeater for the lowest-priority existing talkgroup is reassigned to the emergency call. To accomplish this, the receiving radios on the existing lower-priority call are instructed to terminate that call, and the radios on the emergency call are instructed to tune to the frequency of that voice repeater. Unfortunately, the transmitting radio on the lower-priority call cannot be instructed to terminate the call. This can cause the emergency radio to compete with the lower-priority radio, resulting in distorted audio or no audio.

Telephone Interconnect

The **telephone interconnect** feature allows system users to answer or make calls to telephone users from the user's radio, similar to a cellular phone. The difference between telephone interconnects and a cell phone is that the trunked user cannot transmit and receive simultaneously. Telephone interconnect was a much more valuable feature before the cell phone became commonplace. In addition, similarly to **private call**, it is difficult to predict telephone interconnect usage during system design. This can cause telephone interconnect use to affect operational use of the system.

Dynamic Regrouping

The **dynamic regrouping** feature allows an authorized system administrator to assign a radio to a specific talkgroup remotely. The purpose of this feature is to allow multiple radios to be grouped together on a talkgroup for operational purposes. This feature is limited in function due to the potential delays while the radio is assigned to the new talkgroup; because of this, few agencies use this for critical operations.

Designing a Trunked Radio System

Trunked radio systems are complex combinations of radio equipment with computer control systems and require skilled engineering to design an effective system properly. Trunked radio systems have been in use for over 20 years, and the manufacturers of these systems are fully capable of delivering a system that is technically reliable. These systems are designed and manufactured to be as reliable as conventional radio systems.

The design of the overall system, including the system's coverage and capacity, involves considerable effort to design a communications system that is effective for the community and agencies that will use it. The system must have the capacity to accommodate the needs of all of the users of the system, and must provide useable coverage in all of the agency's service areas. It is critical that the end users are involved in the specification of these parameters.

Capacity Design

The capacity of a trunked radio system is the amount of communications traffic that the system can support in a given amount of time. The frequencies in the trunked radio system are shared among users and assigned to conversations as necessary. If there are more talkgroups (i.e., channels) than there are frequencies, which is usually the case, then the potential exists for calls to be blocked.

It is most desirable for public safety users never to have a call blocked, although this never can be guaranteed in a system with shared frequencies. Manufacturers use statistical models to estimate the traffic presented to the trunked radio system. These models are based on historical traffic information collected from other customers, along with predictions of usage based on experience with similar agencies. This historical information may not represent operations in your agency. In addition, the traffic information may not represent peak loading, but only average loading. If the system is designed and constructed for average loading, and performs as designed with average loads, then it may not be able to provide adequate service when confronted with abnormally high loads. These high loads can occur during natural disasters, or large-scale incidents such as train derailments, plane crashes, or multialarm fires.

An important concept is that all users of a trunked radio system affect the system's performance and can affect other users. The channels used by the system are shared among all users of the system and, like any other shared resource, all users must be aware of their impact on other users and must act accordingly. For example, some users may talk excessively and use the trunking system to discuss issues best discussed face-to-face or on the telephone. Proper user education and the establishment of a formal communications order model can help prevent unnecessary system load.

Coverage Design

If they are to be used for firefighting operations, trunked radio systems must be designed to provide radio coverage inside buildings. System manufacturers estimate what the system will require in terms of radio tower sites and other system components to provide coverage on the street. This becomes the base reference point and is referred to as 0 decibel (dB). In-building coverage levels are dictated by the construction of the buildings from which the users need to communicate. The heavier the construction, the higher the dB level needed to provide radio frequency penetration into the structure.

During system design, the service area is analyzed and geographic areas are categorized based on the structures within them. For example, the central area of a city may have highrise structures that require the highest penetration signal levels. The area surrounding the highrise district may consist of midrise and warehouse structures requiring less signal level to penetrate the structure. In suburban areas, even less signal generally is required to communicate on the interior of a structure. Areas with the greatest radio frequency penetration demand will have a higher number of radio sites than areas with lesser penetration. When a building such as a hospital or school is built in a predominately suburban area, the radio system will not provide in-building communications because the area was designed for residential structures. Interior radio system coverage is dependent on the ability of the system designer to estimate the signal loss accurately for each building type. During testing performed by NIST, building losses as high as 50 dB were found in a 14-story apartment building.[6] The actual radio frequency losses often are much higher than the standard recommendations that system manufacturers use today. This results in marginal in-building communications in many structures.

System manufacturers and designers will never guarantee, and it is impractical to expect, 100 percent coverage. It is impossible to guarantee 100 percent coverage in any city. There is always some corner of a building that a radio system does not cover. The problem with no coverage from a trunked system is very different from a simplex area. In a trunked system, no coverage means no communications.

Since trunked radio systems can have several levels of radio frequency penetration, the users of the system need to be aware that a particular building type in one area of the system may have communications, while the same building in another area may not have communications.

Figure 26 – Coverage Map.

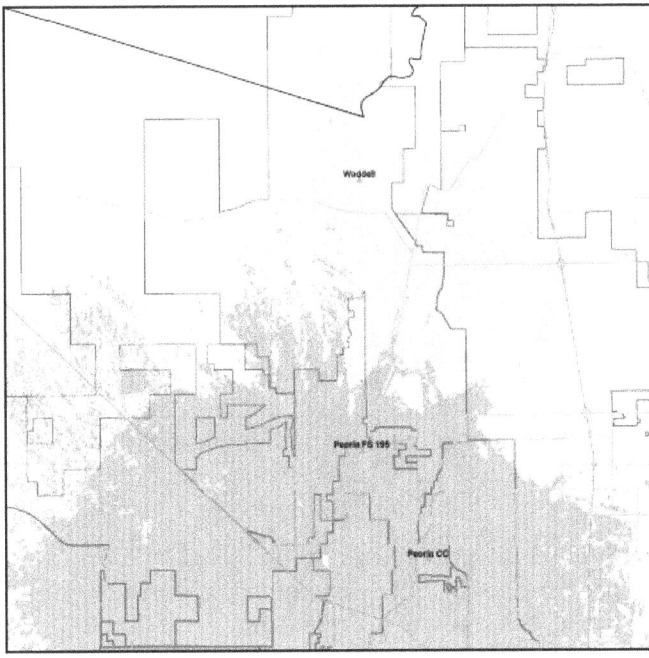

Bidirectional Amplifiers

To overcome trunked system in-building coverage difficulties, BDAs often are used to rebroadcast the trunked system in buildings. BDAs also can be used with conventional duplex radio systems. There are many types of BDAs; all require electrical power and some type of antenna system. Often the antenna systems are installed in the plenum spaces of commercial structures. Most BDA systems include battery backup power to keep it operational if a loss of commercial power occurs.

BDAs work well for incidents such as EMS calls and law enforcement incidents where there is no fire involvement in the building or building systems. In a structure with active fire, the building and building systems are affected directly. The building environment changes with the introduction of fire: Temperatures rise and particulate matter is suspended in the atmosphere. Firefighter actions to eliminate the fire can also have a detrimental effect. As water is applied to the fire, steam is generated and may have an effect on electronic equipment. The moisture mixes with the suspended materials, and acids are formed. These acids can cause intermittent failure of exposed electrical contacts over time. As with all electronics, BDAs are subject to failure when exposed to high heat and moisture. Other actions taken during firefighting operations also could destroy the BDA system. Firefighters checking for extension using pike poles may inadvertently tear the BDA antenna system down, rendering the BDA useless and causing loss of communications inside the building.

Figure 27 – In-Building Communications System.

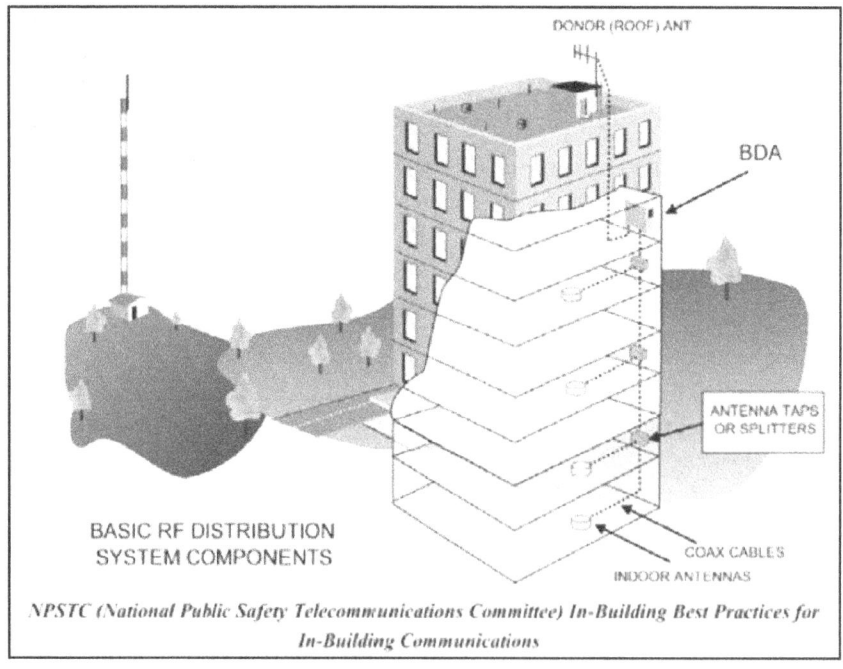

BASIC RF DISTRIBUTION SYSTEM COMPONENTS

NPSTC (National Public Safety Telecommunications Committee) In-Building Best Practices for In-Building Communications

Identifying the buildings that need BDAs and installation of the equipment is a monumental task, especially in fast-growing metropolitan areas. BDAs, like any other transmitters, require periodic maintenance to keep the equipment operating at peak performance. To maintain BDAs in a system requires staffing and technical expertise to keep the equipment operating properly. As building density increases in a given area, a building that did not need a BDA when constructed may need one as it is surrounded by new construction. This requires periodic radio frequency surveys to determine if new BDAs are needed.

Many municipalities have developed fire codes that require installation of this equipment. The codes often require BDAs when a building exceeds some square footage value, during additions increasing square footage by some percentage of the original, or in all-new construction.

Vehicular Repeaters

Some municipalities have recognized the weaknesses in BDA systems and have installed vehicular repeaters (VR) on fire apparatus to provide in-building coverage that is suitable for firefighting operations. A good example of this approach is the Washington, DC, Fire and EMS Department. Each apparatus is equipped with a VR that is activated manually prior to entering the involved structure. These repeaters are operated in the duplex mode, meaning that the users transmit from the portable radio, it is received by the repeater, and then retransmitted to the other portable radios on the fireground.

In the duplex mode of operation, fireground radios are not communicating directly with one another. The radios are dependent on the repeater for communications. If an interior crew encounters an area where they do not have repeater coverage they can switch to a talkaround channel to communicate directly with other interior crews. The other interior crews need to change to the talkaround channel to communicate with the out-of-range crew. When talkaround is used, the unit on talkaround can hear radio traffic through the repeater but cannot transmit to other units unless they also change to talkaround. The talkaround function can cause some confusion unless the unit that switched to talkaround clearly communicates the channel change to other units on the fireground.

Figure 28 – Vehicular Repeater System (Duplex).

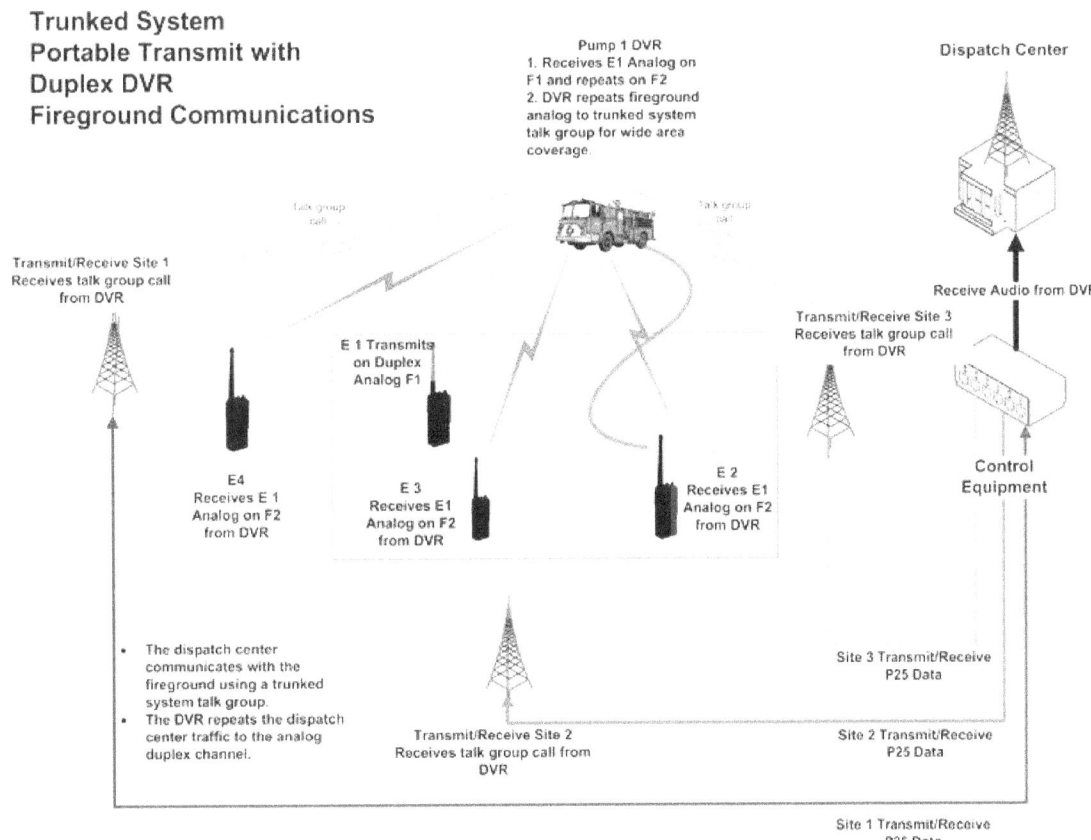

The Phoenix, Arizona, Fire Department is engaged in research on the use of VRs in the simplex mode with analog modulation. This allows crews operating on the fireground and the interior to communicate directly to one another without reliance on any equipment on the exterior of the building. The Phoenix Fire Department has performed extensive testing to ensure that simplex operations maintain positive communications with the fireground, interior crews, and Command.[7] This version of the VR retransmits simplex fireground traffic onto a trunked radio system talkgroup for wide-area communications to the dispatch center and other units that are monitoring. The dispatch center and other units on the trunked system side can communicate with the fireground by transmitting their radio traffic on the talkgroup, and the repeater will retransmit it on the fireground simplex channel.

During the Phoenix Fire Department's testing, the VRs provided predictable levels of in-building communications from the fireground to the dispatch center. Success rates ranged from 96 to 100 percent on the buildings tested.[8] This level of predictability is based on all responding apparatus having a VR unit. In contrast, with ground-based infrastructure, the interior crew may be communicating to a receiver site several miles from the incident, and the communications success rates to the dispatch center will vary based on the building/incident location in relation to the ground-based infrastructure.

Figure 29 – Vehicular Repeater System (Simplex).

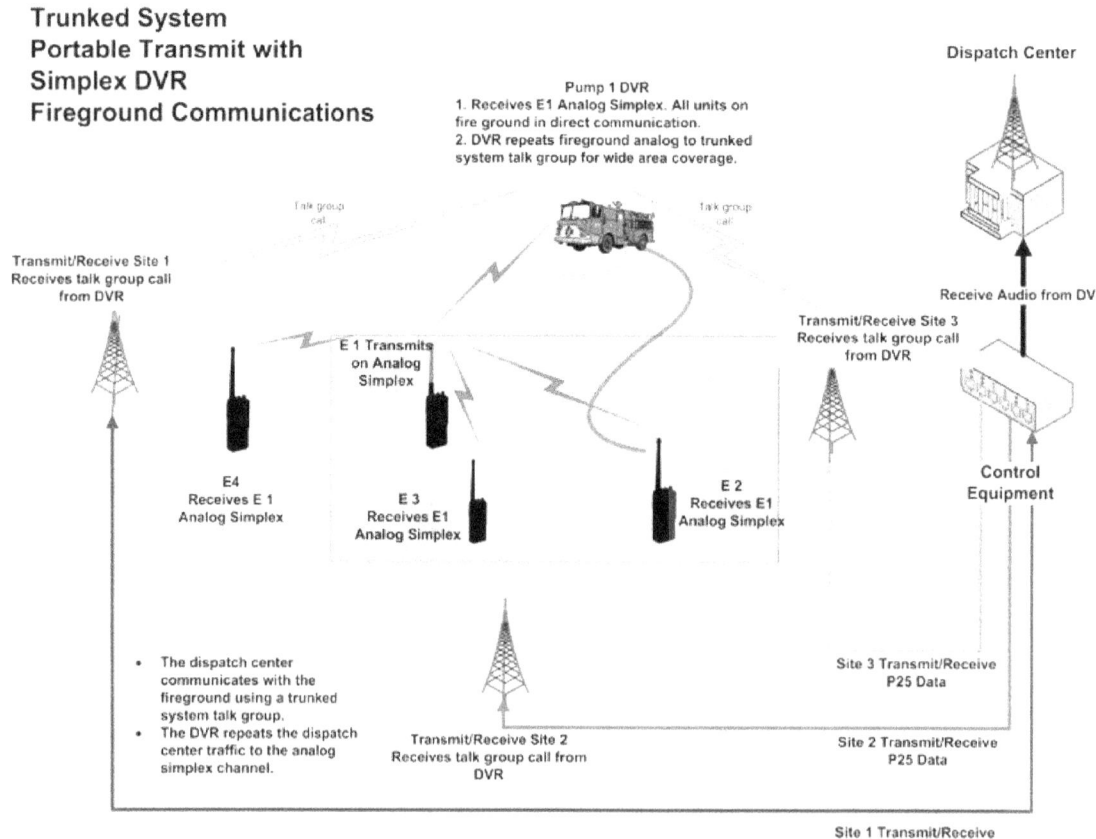

Vehicular repeaters, like any other technology must be operator friendly. The Phoenix Fire Department analyzed the tasks performed by a CO en route to a fire and determined that the activation of the VR must be automatic. The VR is interfaced with the mobile computer terminal (MCT) on the fire apparatus. Upon arrival the CO presses an onscene button to indicate a status change to onscene. If the CO does not press the button, the dispatch center can initiate the status change remotely to activate the VR.

Summary

Trunked radio systems are the most complex of public safety radio systems. They are built on all of the principles and technologies developed in radio systems over the past 50 years. As with all radio systems, the coverage of the trunked radio system is the key to its operation in a way that meets the firefighter's needs. Users can live with systems that lack telephone interconnects, private calling, and paging, but they cannot operate safely inside a hazardous atmosphere with a radio system that does not provide reliable communications.

SECTION 6
SYSTEM DESIGN AND IMPLEMENTATION

Project Organization

Designing and implementing a communications system is an extremely complicated process. It is important to create a structured organization to provide input, carry out decisionmaking, and do implementation work on the project. Get the organization established before beginning the project.

Everyone affected by the fire communications system should have a hand in its selection. This doesn't mean everyone participates at every step, but it does mean stakeholders must be consulted and their needs given serious consideration. If any constituency gets left out of the planning process, those needs may get overlooked, and the result could be a system that fails to meet the requirements and expectations of the entire community.

Stakeholders include

- front-line firefighters and the teams that support them in the field;

- dispatchers and others who provide support away from the scene;

- supervisors and managers at all levels;

- fire department leadership;

- union representatives;

- elected officials; and

- personnel from other agencies that collaborate with the fire service.

In today's world, interoperability is a major concern: The ability of public safety agencies to communicate with each other is critical when events require them to coordinate a joint response. Many localities are answering this need by designing large networks that will be shared by multiple departments and sometimes by multiple cities and counties.

If your community is working on a network that will be shared by other entities in addition to the fire service, you will be collaborating with representatives of those organizations. You'll face the challenge of giving each agency's needs appropriate weight. Often the law enforcement component of the system may drive the overall direction of the project, but it's essential for the fire service to make system designers aware of the needs of the fire service and make sure that the system is designed to accommodate those needs.

Each community has a different approach to organizing the planning effort. Typical efforts include

- A steering committee with top leadership setting the overall policy agenda. Every attempt should be made to have fire department management and labor leadership as participants on the steering committee at the project's earliest stages.

- Working groups that are assigned to complete specific tasks and report back to the steering committee. If you are appointed to one of these groups, it sometimes can be difficult to determine exactly what your role is supposed to be—both as a group and as individuals. It is important that fire department management and labor are involved in establishing the goals and expectations for each work group.

Requirements Definition

You may be collaborating with other departments to build a shared multiagency network or you may go it alone on a network for the fire service only. Either way, the more you learn about your department's needs, the more effectively you can represent the perspective of the fire service in your community.

Design and procurement of radio systems for fire departments are technical and very expensive. Many departments rely on expertise outside of the fire service to advise them on communications technologies. Often these technical experts do not have a complete understanding of the fire service or special requirements related to fireground communications. As a result, many communications systems are built to design parameters based on incomplete or inaccurate information. In support of the design process, fire departments should conduct an analysis to determine the most appropriate radio communications technology. The development of a "Requirements Definition" provides an opportunity to analyze communications needs based on operational practices and inherent risks associated with fire operations. The Requirements Definition also provides a measurable parameter set to evaluate the current radio system.

Sample Requirements Definition

1.0 Functional Requirements:

1.1 Hot Zone Operations

 1.1.1 Provide immediate, uninterrupted, predominately local (simplex direct channel), radio to radio (crew to crew) communications (bidirectional) without changing channels.

 1.1.2 Crews operating radios in a contaminated atmosphere breathing from SCBAs will be required to use the simplex radio channels.

 1.1.3 All radio units operating in the hot zone must be equipped with a PTT ID signaling feature.

 1.1.4 All radio units operating in the hot zone must be equipped with an emergency ID signaling feature.

1.2 On-Scene Incident Operations (Fireground)

 1.2.1 Provides immediate critical, uninterrupted, predominately local (simplex or direct channel), crew-to-IC communication (bidirectional).

 1.2.2 Provides wide-area-type operation for noncritical functional support for 1) Staging, 2) Logistics, 3) rehabilitation, and 4) investigations. IC needs to be equipped for this type of operation.

 1.2.3 All radio units must be equipped with a PTT ID signaling feature.

 1.2.4 All radio units must be equipped with an emergency ID signaling feature.

1.3 Dispatch Center

 1.3.1 Provides uninterrupted, wide area type operation to link the dispatch center to "all" hot zone and onscene incident critical communications (simplex direct channel traffic). Supports reporting from Incident Command (tactical benchmarks) and Command backup, notification to Incident Command of elapsed time and available resources, monitoring and recording of radio traffic, and monitoring of emergency and Mayday traffic.

1.3.2   Fire call (potential hot zone) — The dispatch center will dispatch the call information initially using the wide-area system/channel. The initial call will include the assigned tactical, critical (simplex direct) channel assignments. (See 1.3.4) All further incident communications with the dispatch center radio operator (both transmit and receive) will be heard on the assigned tactical (simplex direct) channels.

1.3.3   Critical radio traffic with the dispatch center (talk-in and talk-out) from the incident should be balanced to ensure consistent communications with units.

1.3.4   On dispatch, responding units should be assigned a wide area system/channel(s) or tactical simplex direct channel(s) based on the nature code of the call. On larger incident(s), functions on the fireground will be classified as critical (C) or noncritical (NC). Critical functions will be assigned a tactical simplex direct channel(s) and noncritical will be assigned a wide area system/channel(s).

1.3.5   Initial incident reporting — Provide a mechanism for the initial crew arriving at an incident to immediately report the incident situation to the dispatch center.

1.3.6   Alerting tones — Provide a mechanism for the dispatch center to generate alerting tones (i.e., emergency traffic) that will be received by all units (crews) operating on the tactical critical simplex channels.

1.3.7   EMS call (nonhot zone) — The dispatch center will dispatch the EMS call information initially using a wide-area system/channel(s). The initial EMS call will include the assigned tactical noncritical wide-area system/channel(s). All further noncritical communications (See 1.3.4) with the dispatch center (both transmit and receive) will occur on the assigned wide area system/channel(s) manned by a Tactical Radio Operator (TRO).

1.3.8   Connectivity — The dispatch center must have continuity of communications with field radio units consistent with typical public safety standards.

1.3.9   Status display — Provide a monitoring system that gives dispatchers a visual indication of the "system status" (green — normal; yellow — site trunking somewhere; red — outage somewhere). Dispatchers need immediate information related to any outage to allow determination of operational impact to fire services. All service-affecting situations for the networks should be immediately reported to the dispatch center.

1.3.10 Capacity indicator — Provide an indication for the dispatchers of system capacity remaining as available for use. Also provide a "busy" indication when a certain level of capacity (grade of service threshold) remains.

1.4 Special Operations

1.4.1   Other special operations in support of fire scene operations, such as use and coordination of a helicopter, may require the use of the simplex direct channels for use on an incident response as assigned for use by the dispatch center during the dispatch process.

1.4.2   Secure communications will be a function of using an appropriate encryption technology.

1.5 Intra-Discipline Operability

1.5.1   The fire department requires radio operability with other regional fire services that operate using different radio communications systems. This is an intradiscipline requirement to support mutual-aid operations.

2.0 Technical Performance

2.1 Analog Modulation

2.1.1   All hot zone and critical onscene incident tactical radio transmissions will use a simplex/ direct analog transmission mode.

2.2 Coverage Area

2.2.1   Primary coverage areas are within the service areas of the fire department.

2.3 Coverage Performance (for the coverage areas defined in Section 2.2)

2.3.1   Coverage performance defined for hot zone and onscene incident critical communications for 1) radio unit to radio unit (crew-to-crew operations), and 2) radio unit (crew)-to-IC is defined as the use of portable radios, operating on the assigned frequencies.

2.3.2   Coverage for onscene incident critical communication for radio units (crew and IC) using the simplex analog communications mode with the dispatch center location is defined as the following: equivalent to DAQ 3.4 / 95 percent area reliability for portable in-building communications (talk-out from the portable to the dispatch center and talk-in from the dispatch center to the portable), in the presence of noise, interference, and other factors as listed in TIA/EIA TSB-88, using the following margins above those required for the defined area reliability and based on the manufacturer's equipment.

> **2.3.2.1**   In the areas considered residential, 12 dB building loss will be added to the baseline signal level required for on street portable coverage (trunking).
>
> **2.3.2.2**   In the areas considered medium density, 17 dB building loss will be added to the baseline signal level required (trunking).
>
> **2.3.3.3**   In the areas considered high density, 23 dB building loss will be added to the baseline signal level required (trunking).

2.3.3   Coverage for tactical noncritical communications is as provided by the wide area system/channel(s).

2.4 Capacity

2.4.1   For each coverage region an adequate number of simplex channels are required, based on appropriate traffic studies.

2.4.2   For the entire coverage area, an adequate number of wide-area channels are required based on appropriate traffic studies.

2.4.3   The system shall be designed to accommodate expansion in both the area-specific channels and the wide-area channels as capacity requirements increase.

3.0 Interoperability

Interoperability is defined as a communication link (connectivity) and the appropriate operating practices between the primary fire department, automatic aid, mutual aid, and corresponding law enforcement entities that allow radio users to communicate with each other on demand and in real time.

3.1   Connectivity between the dispatch center and these agency's corresponding law enforcement entities.

3.2   Connectivity between the dispatch center and other law enforcement entities including sheriffs, highway patrol, and the Federal Bureau of Investigations (FBI).

3.3   Development of operating practices and training for dispatchers and radio operators.

4.0 Network Transport Requirements

4.1   Capacity — The network transport facilities will be expanded to support the critical response and reliability requirements consistent with public safety services.

4.2   Reliability — Any system transport services should use public safety owned and maintained facilities to support critical reliability and maintenance criteria.

5.0 Site and Facility Requirements

5.1   All towers, shelters, and other remote communications infrastructure will be equipped with the appropriate electrical and mechanical facilities to support the critical response and reliability requirements consistent with public safety type services.

Development of Requirements Definitions should consider these standards:

Standards Related to Incident Operations

1. NFPA 1221, 2002 Edition.

1.1   **Section 4.1.2(2)** — In the event of the loss of function of communications equipment, an alternative means of communications shall be readily available.

1.2   **Section 4.1.7** — Equipment … capacities shall be designed to handle peak loads rather than average loads.

1.3   **Section 6.6.1** — Communications centers shall have a logging voice recorder, with one channel for each of the following: (1) Each transmitted or received radio channel or talk group

1.4   **Section 8.1.2.6** — The radio communications system shall be monitored as follows: (1) It shall indicate faults and failures, (2) Audible and visual indications of faults or failures shall be provided to the telecommunicator and radio system manager, (3) Monitoring for integrity of portable radios and radio equipment installed in an Emergency Response Facility (e.g. Fire Station) and in emergency response vehicles shall not be required.

1.5   **Section 8.3.1.3** — A separate simplex radio channel shall be provided for on-scene tactical communications.

1.6   **Section 8.3.4.1.26** — Tactical Communications. Trunked system talkgroups shall not be used to fulfill the requirements for the provision of a simplex radio channel for on-scene tactical communications.

1.7   **Section A.8.3.1.3** — The telecommunicator should have the ability to monitor all tactical radio communications.

2. NFPA 1561, 2002 Edition.

2.1   **Section 3.3.23** — Radio Communications. Definitions of Command Channel, Dispatch Channel and Tactical Channel.

2.2   **Section 4.3** — Communications. Section on communications procedures to support incident management system and operating procedures.

3. OSHA (Act of 1970) — Section 5(a). Duties

(a)(1)   Each employer shall furnish to each of his employees employment and place of employment which are free from recognized hazards that are   causing or are likely to cause death or serious physical harm to his employees

(b)   Each employee shall comply with occupational safety and health standards and all rules, regulations, and orders issued pursuant to this Act which are applicable to his own actions and conduct.

4. OSHA Reference — 29 CFR 1926.65 Hazardous Waste Operations and Emergency Response

(d)(3)   "Elements of the site control program." The site control program shall, at a minimum, include: …, site communications including alerting means for emergencies …

Needs Analysis

You may be collaborating with other departments to build a shared multiagency network, or you may go it alone on a network for the fire service only. Either way, the more you learn about your department's needs the more effectively you can represent the perspective of the fire service in your community.

The planning horizon for a new communications system can range from a few months to several years. Once installed, the system could have a life of 10 years or more. The following are some things that must be considered.

Community needs — Anticipate population growth, density changes, geographic expansion, alliances with other communities, and evolving issues in homeland security and all-hazards management. Any investment you make today should have the potential to grow tomorrow.

Organizational changes — Consider potential staffing changes, departmental realignments, the creation of new work teams and task forces, greater collaboration with State and Federal agencies. Will you be hiring more firefighters, opening or closing stations, or fielding specialized teams such as hazardous materials, weapons of mass destruction (WMD), wildland firefighting, technical rescue, or others?

- Be prepared with statistics that reinforce your department's importance to the community: How many incidents you handle each year, how many citizens receive service each year, and how many lives are saved. These can be hard to quantify, but some research should produce numbers you can use.

- Be familiar with your department's planning initiatives and be prepared to talk about anticipated growth, potential incidents, and disaster scenarios to demonstrate the importance of fire service preparedness.

- Focus on results. It's not a question of how many antenna towers you have, it's whether firefighters can hear to coordinate tasks and strategies or hear emergency traffic or a Mayday call when they're working inside a building. Emphasize how each decision affects the safety of your personnel and citizens.

Also, with the current focus on interoperability, don't lose sight of the basic mission. It is still more important to be able to respond effectively and safely to the everyday incidents than it is to provide for every possible (and unlikely) disaster scenario. This is not to say that interoperability is not important, but don't sacrifice a system you can use for a rarely used feature.

Evaluation of Current System

The development of a Requirements Definition provides the metrics, the measures, to evaluate the current radio system. Identifying gaps or lack of gaps will assist in determining the need for a new system, or if updating the old system is cost effective. During the evaluation what are the gaps in the current system you are using? Consider gaps in functionality, coverage, and in meeting Federal and local government mandates (e.g., narrowbanding).

Take a Snapshot of Your Communications Today

What is the current state of your fire communications? This is not an easy question to answer. It's not uncommon for a department to use more than one communications system and, even with the same equipment, procedures can vary markedly. Collecting this information and pulling it all together in one place is a necessary step and one that requires the commitment of time and resources. Few departments keep statistics about radio usage and performance, so you'll have to generate much of this information from scratch. Many departments bring in a consultant at this phase, especially to help with the more technical aspects of the job, such as charting call traffic and measuring grade of service. A consultant also can be helpful in collecting "softer" data; sometimes it's easier for an outsider to interview users and get their honest opinions.

Evaluation of Proposed Technologies

After collecting the statistics of the current communications system, and armed with a Requirements Definition, a comprehensive evaluation of your current system and proposed technologies can be made. The Requirements Definition becomes the scorecard where the current and proposed technologies can be graded on compliance, partial compliance, or noncompliance. All components of the Requirements Definition may not be compliant in all technologies. Each department will have to evaluate each component of the Requirements Definition and derive an importance factor to determine if noncompliance or partial compliance is acceptable for their department.

Create a User Group

The group should be composed of actual communications users who perform firefighting and support tasks. Avoid having technical and support personnel who are not performing actual firefighting and rescue tasks, as this often confuses the user's message. This group can help determine the what, why, where, when, and how. Ask what is good about the current communications system and what changes would help the users perform more efficiently or effectively. Use this information as part of the overall plan. Compile a complete and accurate picture of how the fire service in your community communicates today before you can get on with the process of making it better for tomorrow.

Technical Options and Conceptual Design

What technology is available to close the gaps between operational needs, Federal, State and local mandates, and the current system? Select the best combination of technologies that close the gaps without compromising the mission. Keep in mind the safety of firefighters, mission effectiveness, and long-term sustainability.

New technologies — While you can't predict every future capability, you can read news reports and technology journals for emerging systems, pilot programs, and development projects. Look for military spinoffs that will be adapted to the fire service. This how we got thermal imaging cameras for locating fire victims and missing personnel, global positioning system (GPS) location systems, and radios that can operate using different frequency bands and protocols as needed. In the next few years, radio networks will be able to support a range of new features. Even if you don't have the funding to activate these features today, you may choose to invest in a system that will be capable of supporting them later. These may include

- Voice-activated intercom systems that would allow multiple interior attack firefighters to communicate while keeping their hands free.

- Large, accessible buttons on turnout gear to enable immediate distress signaling.

- Radio-linked PASS devices that alert a Safety Officer if a firefighter remains motionless for too long.

As an advocate for the fire service, you can use these tips to help ensure that your concerns will not be lost in the shuffle. While many of these technological improvements will prove to be beneficial to firefighters in the future, this guide is directed primarily at voice communications.

Should You Hire a Consultant?

Time, staffing, and know-how are factors in deciding whether to hire a consultant. Do you have people with the necessary technical capabilities and an understanding of complex modern communications systems? Does your organization have time to do the job alone? Can you obtain the necessary staff internally? Do your people know how to perform the assigned tasks? If the answer is "no" to any of these questions, consider getting the assistance of a consultant.

Even if you have some degree of technical capability in-house, the use of an outside consultant brings the benefit of experience. The consultant has (or should have) more experience than you in dealing with communications challenges and providing communications project oversight. The consultant also provides an outsider's fresh viewpoint, which can be valuable. A consultant can be hired to perform a single, clearly-defined task, or to take on a more comprehensive role. Often, it's wise to hire a new consultant for a small-scale project and see how it works out before turning over a large-scale responsibility.

If you decide to use a consultant, ask these questions before you hire:

- Have you worked with other public safety agencies before?

- Have you worked with fire departments before?

- Have you worked with fire departments of our size?

- Are you able to provide assistance to overcome budget issues, such as grant writing, understanding the bond process, and creative financing solutions?

- What types of systems have resulted from your work?

- What are some of your successes, and what were some of your failures and how did you overcome them?

- Who are your references and how can we contact them?

- Investigate possible relationships between the consultant and vendors.

Where to Get Advice

Whether or not you use a consultant, investigate these alternative sources of assistance and information:

Other communities — Chances are that another town near you has been through this process already. Look at other departments of comparable size and contact their committee members and arrange a meeting or conference call where you can "pick their brains."

Conferences — Attend fire and public safety conferences with an eye for communications sources. Programs, panels, vendor displays, demo projects… they're all good places to get information and hook up with others who have experience they're willing to share.

Vendors — Manufacturers and system integrators often can provide brochures, white papers, and similar information resources. This is another place to find information about technical issues. An established vendor understands that well-informed customers are the best customers and that providing accurate information is one way to build a strong, lasting relationship and ensure the customer's long-term satisfaction. Be cautious of vendors who are in business solely to make money, not necessarily to meet your needs. Currently there is a lack of real competition due to the extremely small number of companies who build these systems. You must have a strong labor/management commitment not to use a system until it is proven to be safe and cost effective and to get the best system performance from the contractor.

Government and professional organizations — Several national organizations act as clearing houses for information about public safety communications. Again, a word of caution: While the organizations listed below do good work in the areas of interoperability and system standardization, no other organization outside of the IAFF is focused on the special needs of firefighters involved in interior operations.

U.S. Department of Homeland Security (DHS) SAFECOM program — The SAFECOM program's mission is to help local, tribal, State, and Federal public safety agencies improve response though more effective and efficient interoperable communications. SAFECOM provides guidance, tools, and templates on communications-related issues and supports research and testing of communications products for public safety. Visit www. safecomprogram.gov or call 866-969-SAFW.

National Interagency Fire Center — The U.S. Departments of Agriculture and the Interior provide information on the use of radios in fighting wildland fires. Much of this information also applies to communications on structural fires. The information includes portable and mobile radio testing results, including digital radios, and training on various topics. For additional information, please see www.fireradios.net

NPSTC — This is a federation of Federal, State, and local associations and agencies. It serves as a liaison among the FCC, Congress, and appointed officials involved in public safety communications. NPSTC was originally formed to implement the recommendations of the Public Safety Wireless Advisory Committee (PSWAC). NPSTC has taken on a wide range of activities related to spectrum policy coordination and the development of new technologies. More information is available online at www.npstc.org or by calling 866-807-4755.

APCO — APCO is a professional organization whose mission "…provides leadership; influences public safety communications decisions of government and industry; promotes professional development; and, fosters the development of technology for the benefit of the public." APCO sponsors the P25 digital radio standards process. APCO's focus is primarily on technical and operational standards relating to communications systems and communications centers. For additional information on APCO, go to www.apco911.org or call 888-272-6911

Funding

After the needs are identified and a technical solution is decided upon, the budget and implementation timeline can be developed. If the budget is developed too early, the system design may be unduly constrained. When this happens, it is inevitable that functionality and performance will be lost. Once the budget is set, it will be very difficult to get additional funding later to "get it right," especially if other agencies are pushing forward.

Alternative Funding Sources

Funding is a huge issue, but it should not be your first consideration when assessing your communications requirements. With the renewed focus on public safety and first-response capabilities, more funding is becoming available through Federal, State, and regional government grants. Examples include

- State Homeland Security Grant Program (SHSGP), administered by DHS through the Office for Domestic Preparedness (ODP);

- Assistance to Firefighters Grant Program (FIRE Act), administered by the Federal Emergency Management Agency (FEMA) (www.firegrantsupport.com); and

- Urban Area Security Initiative (UASI) administered by DHS.

In some cases it may be feasible to participate in joint investments with other agencies or nearby communities. This will allow for networking facilities such as core systems, repeater systems, fire-alerting systems, and towers. Costs can be shared among several different organizations. This also improves day to day interoperability among these organizations.

Explore leasing agreements and other financing alternatives to up-front capital investment. Phased implementation plans and adaptable networks that start small and add more capabilities over time as the funding becomes available are also an option. Do not allow cost to become a barrier that prevents your community from building the fire communications system its citizens and your colleagues deserve.

Grant Writing

If you decide to try to obtain funding through a grant, it is important that you get started on the grant proposal early and spend the necessary time to get the proposal right. A successful grant proposal is well-prepared, thoughtfully planned, and concisely packaged.

Become intimately familiar with the grant criteria and the eligibility requirements. You must be able and willing to meet these requirements. You might find that eligibility would require providing services otherwise unintended, such as working with particular client groups or involving specific institutions. You may need to modify your concept to fit. Talk to the grant information contact person to determine whether funding is still available, what the applicable deadlines are, and what process the agency will use for accepting applications.

Determine whether any similar proposals have been considered already in your locality or State. Check with legislators and area government agencies and related public and private agencies that currently may have grant awards or contracts to do similar work. If a similar program already exists, you may need to reconsider submitting the proposed project, particularly if duplication of effort may be perceived.

Enlist the support of community leaders. Once you have developed your proposal summary, look for individuals or groups representing academic, political, professional, and lay organizations that may be willing to support the proposal in writing. The type and caliber of community support is critical to your proposal's ability to survive the initial and subsequent review phases.

You probably can develop the proposal without hiring a grant writer. Most fire grant programs are designed so that an astute member of any fire department can write a successful application. FEMA has a help desk staffed with competent professionals who help applicants through the process. In addition to the help desk, FEMA offers free grant-writing seminars and supports a Web site with helpful grant information (www.firegrantsupport.com). Additional information that may assist grant writers is available through the Responder Knowledge Base (www.rkb.us).

Procurement

Based on the identified funding and the conceptual design, solicit companies to construct the system. The procurement process typically will involve the development of a request for proposals (RFP).

Developing the Request for Proposals

The more information about your community, your department, and your needs that you write into an RFP, the better. Vendors need to know about your operational needs and your current systems so they can propose appropriate solutions. If it's not in the RFP, you can't expect to have it addressed properly in the proposals. Use the labor/management process to document user requirements for operations as the foundation for all of the designs and studies that will follow. This is not about technologists and engineers telling you what technology you need. This is about you telling local government leaders and vendors what you need in the field.

It is very important to involve the agency's purchasing personnel early in the purchasing process. This helps ensure that all State and local purchasing requirements are followed, and that important contract language is included in the RFP. The RFP should include a summary of all of the steps taken to get to the RFP stage, including the results of the requirements gathering and current system analysis. The more background information you can provide to potential bidders, the closer their proposals will match you needs. In addition, by removing uncertainty from the purchasing process, you reduce the bidder's risk, hopefully reducing the overall price.

The RFP development stage is a good time to have a consultant involved in reviewing the requirements, and possibly to assist in the preparation of the RFP itself. Much of the RFP can be tedious to develop, and selecting a consultant who has done this work before will reduce the burden on the agency members. In addition, consultants have encountered similar proposals frequently and can include relevant information and experiences into the proposal.

Evaluating Request for Proposals Responses

Modern radio networks employ many different technologies. The best choice for your community usually boils down to striking the right balance between initial cost and long-term capabilities. You need a system that fits your needs and available resources today, with the potential to grow and add more capabilities tomorrow.

The vendor responses to your RFP not only should detail the type of system they're proposing, but also explain why they're recommending it over the alternatives. The vendor should be ready to answer any questions you have about the reasoning behind the recommended system design. Be sure vendors are recommending this design because it best meets your specific requirements.

Some questions to ask about the proposed system and equipment:

- Does the system cover your regular and automatic/mutual-aid service area?

- What is the vendor's solution for fireground communications where the network doesn't provide 100 percent coverage? What will users do if they are outside the range of your network system or indoors where signals don't penetrate?

- Does the system have enough capacity to handle normal and abnormal incidents? What happens if the system becomes overloaded?

- How do other public safety and nonpublic safety users affect the fire department's use of the system?

- How will the system facilitate interoperability with communications systems used by the departments with whom you have mutual-aid agreements?

- How will it alert units of dispatches in fire stations and when out of the station? Can the system accommodate any paging needs?

- Fire-capable end-user equipment (submersible, etc.) is more costly than the radios commonly recommended for police departments; be sure the quote includes the right equipment.

- Are the accessories you need included with each radio (battery charger, speaker microphones, etc.)?

Also, look for an understanding that deploying a new network is not just a technical challenge, but also a major organizational change that requires a full support structure. The vendor's response should include

- Clear identification of how the technology will support your operations and not affect them negatively. Radio systems should be designed and implemented to support your work, not vice-versa. Your existing internal procedures should not be affected negatively by the new system.

- A phased rollout plan for gradual transition from your current system to the new one.

- An upgrade/migration plan for making further changes in the future.

- User training information, including before, during, and after implementation. This is far more important than most people realize.

- System testing and acceptance procedures.

- Practice session information.

- Life-cycle maintenance, network performance monitoring, and repair procedures.

- Software upgrades for radios and the system infrastructure. To evaluate the solution proposed by each vendor, you'll need to understand the relative advantages of the technological choices they are recommending. The next chapter will help.

Implementation

Involve the right people throughout the implementation process. Thoroughly test the system as it is built to ensure that it is meeting needs and expectations.

Successful implementation/integration requires careful attention from the beginning to design compatible links and then test, test, and test again. The vendor's engineers must have a detailed plan that identifies all of the systems to be integrated and defines which capabilities will be made to work together and when. The plan also should include schedules and priorities, and whether the new network will be made operational before all of the integration is completed.

Encourage everyone to ask questions and make comments. You will want to handle concerns and objections early, before they have the chance to evolve into rumors and long-standing gripes.

Before the contract is signed, ask the vendor or consultant to explain the following, and begin to share that information with the rest of your department:

- What operational differences will our users notice between our current system and the new one? How will their procedures change? What new features will be available? Which, if any, of the old features will change or become unavailable?

- What's different for the dispatchers? For field supervisors? For personnel back at the station? For personnel using the in-vehicle radios? For administrators and network managers?

- Will users still be able to use their old equipment, or will they be required to learn new equipment?

- What successes and pitfalls have been experienced by other fire departments implementing this type of system? What have you learned from previous deployments?

Training and Transition

Ensure that all firefighters and Command Staff members train with the system often prior to final switch-over. Inadequate training is an especially critical problem and could endanger the lives of firefighters and the citizens they protect.

Training is far more than simply knowing how to turn on the radio and which buttons to press. Training must not become a one-time experience; firefighters need initial exposure, formal training, and opportunities to incorporate radio usage into other training and simulation exercises. The integration plan also may cover interoperability with systems in other departments or jurisdictions. Interoperable communications must be tested with the joint cooperation of these other agencies, and perhaps their system vendors as well. Training can be broken down into phases, as described below, that lead from general information on the system to specific operational aspects of the system, and finally to periodic refresher training.

Awareness — This phase provides general information. A series of videos, using a live spokesperson, explains what's different about the new system and expectations for the new equipment. The goal is to create interest, not to provide detailed information, and hopefully begin to create champions within your system.

Education — Additional videos are distributed to provide more detailed information on topics such as how to use your radio and what are the direct operational implications of the new system or subscriber equipment. The videos may be broadcast over the department's video network or local cable public safety access channel and also can be available in the station for firefighters to view at will; lesson plans are available on the department's Web site.

Training — Six months to one year before the system's operational deployment, use of the new radios is integrated into fireground training scenarios and in-building tactical preplan surveys. Training is structured in a 3-month cycle. The first month, trainees focus on how to use the radio. In the second month, there's a walk-through. In the third month, the radios are part of a live drill complete with smoke, while trainees wear full turnouts. After this 3-month cycle is completed, a new lesson plan is used in the next quarter, and the cycle continues until the entire set of training classes has been completed.

Transition — By the time the network is operational and transition begins, users will have had 6 months to 1 year of hands-on training. Two-thirds of the total training time is hands-on. Mobile radio training takes place at the time of installation of the equipment in the truck.

Refreshers — Quarterly refresher training (with an emphasis on lessons learned) and just-in-time updates should continue to be given, as well as an annual refresher on fireground communications.

Beyond this training program, which was designed to support the rollout of the new radios, there are implications for other training organizations and curricula. Communications training must be integrated into all phases of recruit training and company training programs:

• Recruit training should incorporate radios from the beginning. In the past, radios were not used during recruit training and a rookie's first day on the job was the first day he or she was given a radio.

• The engineer's academy, captain's academy, and Command Officer's academy, as well as special team training should integrate radio communication throughout the curriculum.

Lessons Learned and Feedback

During the first few months after the initial cut-over to a new system, collect and analyze information regularly on the operation of the system. Share this information among all members of the implementation team and, if issues are found that affect operations, share that with the field users.

All members must be involved in providing feedback on system issues and must be kept involved with the solutions. Get buy-in from the system operator and technical staff to take field user input seriously. Encourage all members to report perceived deficiencies in the system and follow up with the users with updates on their reports. If it appears to the users that their feedback is not acted on, they will stop providing that feedback. It is important to ensure that management is honest with users about the operation and safety of the system. If something isn't working properly, disclose it and find a work-around until the solution is found and in place.

Operation and Maintenance

Ensure that adequate funding is allocated to the operation and maintenance of the system. Just like fire apparatus, the system must be maintained and equipment must be replaced as it becomes unable to serve the agency's needs. Continuously solicit feedback to keep on top of any problems that come up with the system over time.

Throughout the life of the new network, fire service representatives will need a way to handle such ongoing responsibilities as:

- answering users' questions and helping them solve problems;

- incorporating radio usage into new training programs and exercises, and presenting refresher courses;

- monitoring the performance of the system and collecting reports of problems, such as buildings that lack coverage, or situations in which there were not enough channels or talkgroups available; and

- implementing network interoperability links to support new mutual-aid agreements with other communities.

Summary

Developing and implementing a new communications system can be a complex and expensive project. In the case of a large system, it may be the most expensive and most complex project a department has ever undertaken. These facts make it critical that the project is managed adequately.

Establish a project team that includes fire department management and labor representation early in the project lifetime, involving all stakeholders, and ensure that they continue to participate in the implementation process. Gather information on the communications needs of field personnel and compare this to the radio systems they use. This comparison will result in a gap analysis that shows the deficiencies in the current system. The current system description along with the gap analysis can be used to produce a specification for the new radio system.

After the specification is established, a budget can be developed using the requirements and cost estimates developed from similar systems or through talks with potential vendors. Be cautious in reducing the system functionality if the budget is determined to be too large. Removing coverage or features from the system to reduce cost could affect the usability or safety of the entire system.

Once the implementation of the system has begun, familiarization and training should start as well. Early, simple training will provide end users with information on the system in a more controlled manner. If users don't get the information they are seeking, they will find it through another path, or will develop their own.

After the new system has been placed into operation, it is critical to follow up with end users on the operation of the system. Over time, users will find design, implementation, and performance issues with the system that were not discovered prior to cut-over, or that occurred after cut-over. Timely resolution of these issues will ensure that your successful project remains successful in the eyes of its users.

SECTION 7
INTEROPERABILITY

This definition of interoperability is taken from the DHS SAFECOM project:

> In general, interoperability refers to the ability of emergency responders to work seamlessly with other systems or products without any special effort. Wireless communications interoperability specifically refers to the ability of emergency response officials to share information via voice and data signals on demand, in real time, when needed, and as authorized.

Day-to-Day

Most interoperability efforts are driven by the need to meet day-to-day operational requirements. In many large urban areas, the responsible fire department may not require day-to-day interoperability, while some departments interoperate on a daily basis. Since September 11, 2001, there has been significant attention toward efforts to expand interoperability past the day-to-day needs of a public safety agency to address extraordinary events and incidents. Interoperability is required and necessary in today's world. Where and how it happens is based on a logical analysis of operational practices and requirements.

Many fire departments have interoperability with other fire departments. Interoperability between agencies in the same discipline is intradiscipline interoperability. Interdiscipline interoperability is between different disciplines. Intradiscipline interoperability is the easiest to achieve, since there is a common language, terminology, and tactical objectives. Interdiscipline interoperability may not share common terminology or have the same tactical objectives. These factors should be considered in determining where interoperability occurs in the Command structure. A prime example is when law enforcement responds to a house fire for traffic control. Each discipline has very different tactical objectives. As the fire responders fight the fire using the "common language" of the fire service, this terminology may not be understood by the law enforcement component. In addition understanding when to talk and when not to talk becomes a safety issue. In these situations interoperability may be face-to-face coordination with the Command element, or coordination at the dispatch center level. In the example below both disciplines respond to a motor vehicle crash — fire/EMS for medical care and law enforcement for traffic investigation and traffic control.

Figure 30 – Response to a Motor Vehicle Crash.

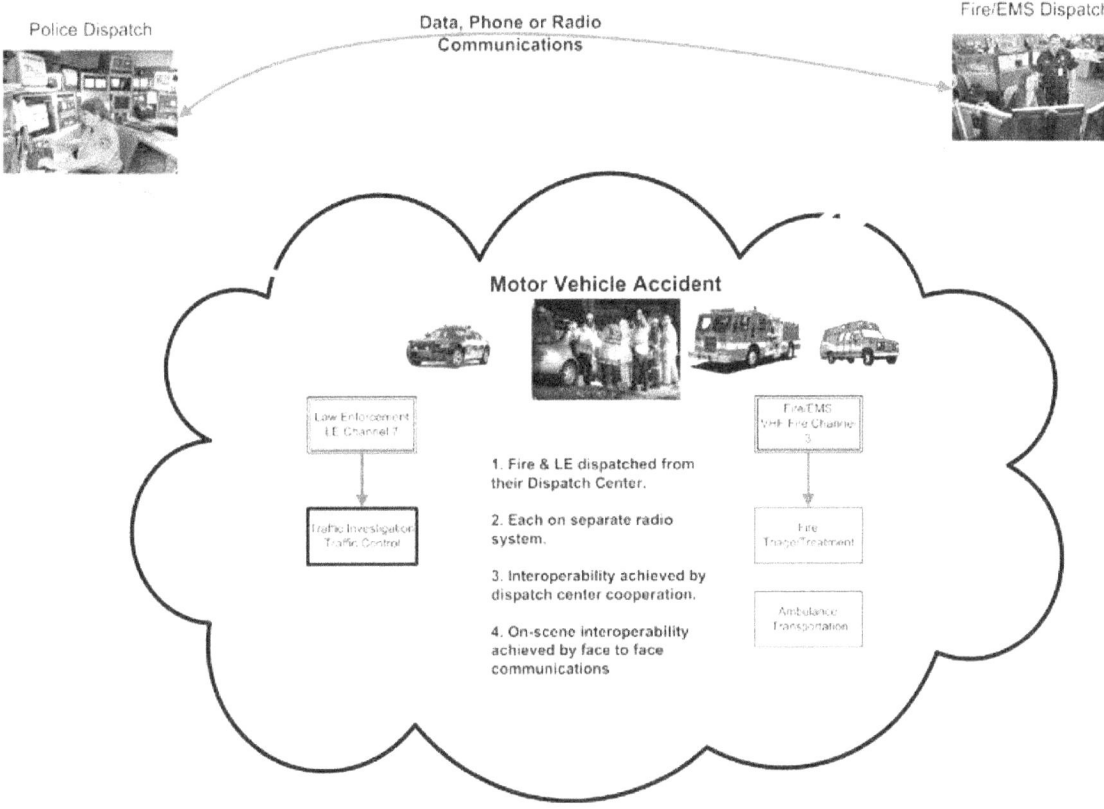

The respective dispatch centers send the appropriate response for each discipline on separate radio channels and maybe even on different systems. Each responds, and when onscene they coordinate at the task level face-to-face. If a shared dispatch center and radio system were used, both units could be assigned on a common channel. SAFECOM would consider this a high level of interoperability.

Large Incidents

As incidents grow, interoperability should be planned for in the Command structure. When developing interoperable Command structures, many interoperability tools may be employed. Technical staff plays a pivotal role in providing these technology tools to meet the operational requirements. The technical staff must be familiar with the operational objectives and Command structure to supply the appropriate technological tools. NFPA 1221 (7.4.10) recommends the use of a Communications Officer at all major incidents, and a Communications Unit Leader is part of the National Incident Management System (NIMS) Command structure. The technical staff should receive the appropriate training to fulfill these roles successfully. Communications Unit Leaders in the NIMS Command structure provide a central point of contact to develop a communications plan to meet the interoperability needs on a large incident. The example below is a large Command structure where multiple technologies are employed to achieve the appropriate level of interoperability for the incident. When interoperating, determining the number of channels needed to support the incident must be a consideration. It is always important to account for the amount of radio traffic on a channel and to reserve some air time for unforeseen needs such as a Mayday. Complex operations that are communications intensive should have their own channel to ensure that there is adequate on-air time and reserve capacity for unforeseen events. Shared or patched channels can be used when there are common tactical objectives. Before patching channels or using gateways that essentially tie channels together, the amount on each of the channels must be a considered. If both of the channels are near saturation the patch or gateway will make communications nearly impossible. Below is an example of a large-scale multidiscipline Command structure where multiple technologies are used to achieve interoperability.

Figure 31 – Incident Communications.

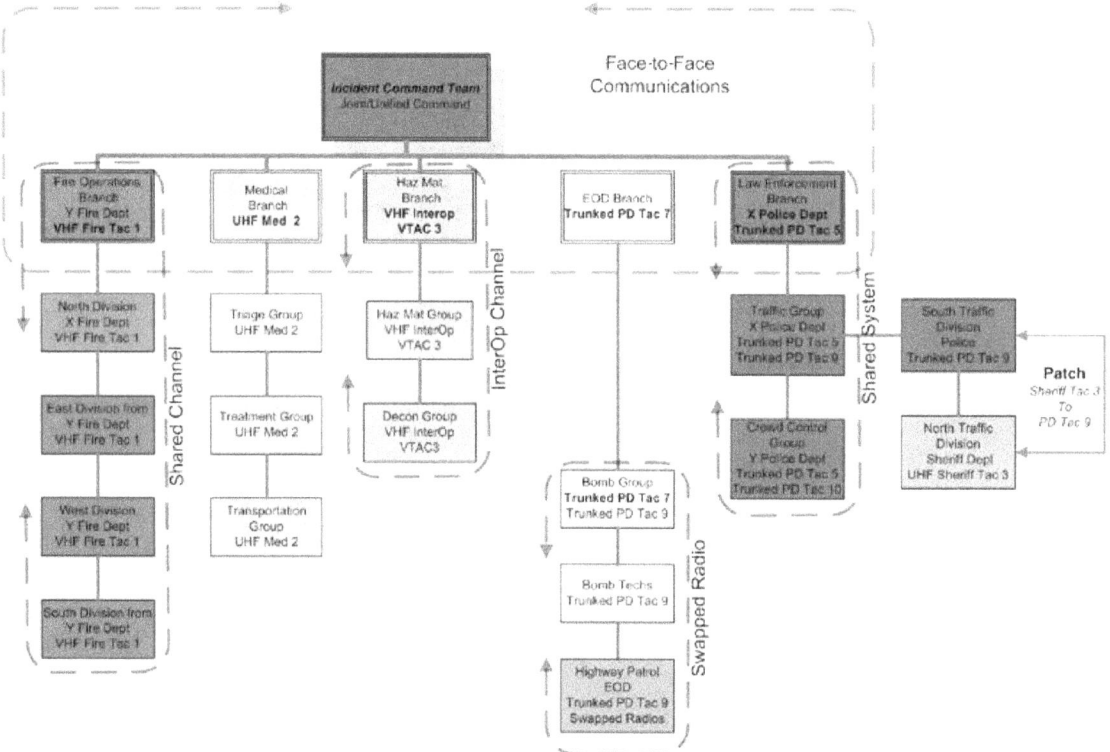

Many technologies are available to achieve interoperability, and often the simplest solutions are overlooked in favor of complex technological solutions. The simple solutions usually are the quickest to implement and easiest to understand. In some instances, face-to-face communications may provide the desired level of interoperability, while in other cases other methods may be necessary. In the example a joint Command structure in a common location allowed the use of face-to-face communications for coordination. When a common Command location is not employed, a strategic-level Command channel is an option.

Many technologies are used to achieve interoperability, and many other factors have an impact on interoperability. SAFECOM is a program within DHS that is tasked with achieving communications interoperability for local, tribal, State, and Federal emergency response agencies. SAFECOM has many documents available that will guide and assist in achieving interoperability. SAFECOM documentation is available at: www.safecomprogram.gov/SAFECOM.

Summary

SAFECOM defines interoperability as the ability of emergency responders to work seamlessly with other systems or products without any special effort. Wireless communications interoperability specifically refers to the ability of emergency response officials to share information via voice and data signals on demand, in real time, when needed, and as authorized. Some interoperability is achieved by coordination at the dispatch level. A high level of interoperability would be a shared radio system and dispatch center that could dispatch multi-discipline responses on a single channel. Interoperability can be intradiscipline or interdiscipline:

- Intradiscipline:
  - like disciplines;
  - common tactical objectives;
  - same language and terminology; and
  - usually easiest to achieve.
- Interdiscipline:
  - different disciplines;
  - different tactical objectives; and
  - different terminology.

NFPA 1221 and NIMS identify the need for a communications coordination officer on large incidents. Proper training is required for technical staff to understand the operational needs on large incidents. As incidents grow, the Communications Unit Leader in a NIMS Command organization is the central point of contact for communications needs and coordination. Many technologies are available to achieve interoperability. Often the simplest solutions are overlooked in favor of complex technical ones. The simplest solutions, such as face-to-face communications and swapping radios, are easy to understand and the quickest to implement. SAFECOM is a program in DHS tasked with achieving interoperability at all levels and is a valuable resource for information.

SECTION 8
RADIO SPECTRUM LICENSING AND
THE FEDERAL COMMUNICATIONS COMMISSION

The FCC is an independent agency of the U.S. Government established by the Communications Act of 1934. It is made up of seven bureaus that are responsible for various communications areas, organized by function. The bureau that is most involved in public safety issues is the Public Safety and Homeland Security Bureau (PSHSB), which

> ...is responsible for developing, recommending, and administering the agency's policies pertaining to public safety communications issues. These policies include 911 and E911; operability and interoperability of public safety communications; communications infrastructure protection and disaster response; and network security and reliability. The Bureau also serves as a clearinghouse for public safety communications information and takes the lead on emergency response issues.

As this description implies, the PSHSB is responsible for rulemaking, licensing, education, and outreach to public safety agencies. Portions of the activities of the PSHSB were previously carried out by the Wireless Telecommunications Bureau, particularly the rulemaking and licensing functions. The outreach and coordination functions were added to create a single bureau to handle all public safety issues.

The rules established by the FCC are located in the Code of Federal Regulations (CFR) Title 47. The section of these regulations that applies directly to land mobile radio systems used by public safety are located in Part 90 of 47 CFR. The Part 90 rules govern the operation of radio systems, as well as the frequencies available for use, what types of agencies are eligible to use the frequencies, and the procedures for licensing the frequencies.

Rulemaking

When the FCC believes that a change is needed to the rules, generally it will first issue a Notice of Inquiry (NOI) asking for general information on the issues related to the change. Next, the commission will issue a Notice of Proposed Rulemaking (NPRM) outlining the proposed rule change. The NPRM allows the public to comment on the proposed change and proposed modifications. After the FCC reviews the comments and proposals, it may issue one or more Reports and Orders (R&O) that provide the final details on the rule changes. This process may repeat as necessary to refine the rule change. In addition, a type of appeals process is allowed, known as a Petition for Reconsideration. During the process, public presentations, comment documents, and expert testimony is heard by the FCC. Fire departments and professional organizations may participate in all portions of the process.

Licensing

The FCC also governs the licensing of radio frequencies to agencies, and this process is handled separately from the rulemaking process, although issues that arise during the licensing process may result in future rule changes.

The licensing process starts with the agency defining the requirements for communications systems, including the type of radio system, the frequency band needed, the number of users that will used the proposed system, and the number of frequencies or frequency pairs required. After the requirements are defined, the agency finds the specific frequencies through a frequency search conducted by the agency, a consultant, or a manufacturer. The FCC Web site has tools to help agencies search for frequencies, including the Universal Licensing System (ULS) which is used to search for existing licenses, as well as for processing applications. The ULS also can be used to search for other agency licenses for examples on preparing a new license. Specific design parameters will be required to license the frequencies, including the transmitter locations, tower height, antenna height, and transmitter power output. Transmitter power output must be specified as "Effective Radiated Power," which increases the actual power output from the transmitter by a gain factor specific to the antenna used in the system.

After all the system parameters and frequencies are determined, an application for license is prepared and sent to a frequency coordinator. The two coordinators used by most fire departments for frequency coordination are APCO and the International Municipal Signal Association (IMSA). The frequency coordinator acts as a prelicense verification to ensure that licenses meet the FCC rules and State and local coordination requirements before being submitted to the FCC. The coordinators take into account interference that may occur to other systems using the requested frequencies and adjacent frequencies. If there are any problems with the application, the coordinator works with the applicant to resolve them.

Once the application passes the frequency coordination process, the application is submitted to the FCC through the automated ULS. Agencies can enter the license information into the ULS and track it as it proceeds. The FCC uses a computer system to perform automated checks on the license and then will assign the license request to an examiner who will perform more extensive checks on the details of the application. The examiner then will either grant the license, or return it to the applicant for modification or additional documentation. If the request does not conform to FCC rules, it may be rejected outright, and will require a reapplication.

If the application does not conform to the FCC rules in Part 90, the applicant may request a waiver of the rules. The waiver process is complicated, and waivers are not granted frequently. An example of a waiver that has been granted is the use of UHF TV channels 14 and 16 for public safety use in the New York and Los Angeles metropolitan areas. These areas had significant needs for additional frequencies in the 1980s, before the 700 MHz and 800 MHz public safety bands were established. The agencies involved presented the needs along with extensive documentation on why the need could not be fulfilled with current frequency allocations. Departments that wish to pursue a waiver must present a detailed, well-thought-out case to be successful.

Federal Communications Commission Actions to Increase Public Safety Spectrum

Historically, all public safety systems used frequencies in the VHF low, VHF high, and UHF bands, with the systems progressing to higher frequencies as technology improved. Most fire and police departments in the U.S. still use radio systems in the VHF and UHF bands and have no plans to move to other bands. However, the population growth in large metropolitan areas has created rising demand for more radio frequencies. In many areas of the country, all available VHF and UHF frequencies are assigned to agencies, leaving no space for growth. The FCC, working with equipment manufacturers and public safety communications organizations, has developed several programs to increase the available frequencies for public safety communications.

National Public Safety Planning Advisory Committee

The first major expansion of frequencies allocated to public safety took place in 1986, when the FCC created the National Public Safety Planning Advisory Committee (NPSPAC) to develop frequency allocations on the 800 MHz band. Prior to the NPSPAC process, public-safety-licensed frequencies in the 800 MHz band were combined with commercial business and cellular companies, and the available frequencies were very limited. The NPSPAC frequencies were put under the control of 55 Regional Planning Committees (RPC). The RPCs are responsible for creating regional frequency plans that take into account agency needs, including metropolitan, rural, and statewide, and are responsible for initial coordination of applications.

Rebanding Below 512 MHz

The NPSPAC process provided additional frequency spectrum for new systems operating in the 800 MHz band, but most fire and police departments in the U.S. still operate in the VHF or UHF bands. To increase the available frequency spectrum for public safety in the VHF and UHF bands, the FCC began investigation into narrowing the bandwidth for frequencies in this band.

In these bands, channels were spaced 15 kHz apart, with transmitters operating with 25 kHz bandwidth. In addition, as shown in the illustration below, adjacent transmitters were separated geographically to minimize interference. It became apparent that, as the population served by these departments grew, their spectrum needs would grow as well, and the existing band plan would become inadequate for the needs.

Figure 32 – Before Narrowbanding.

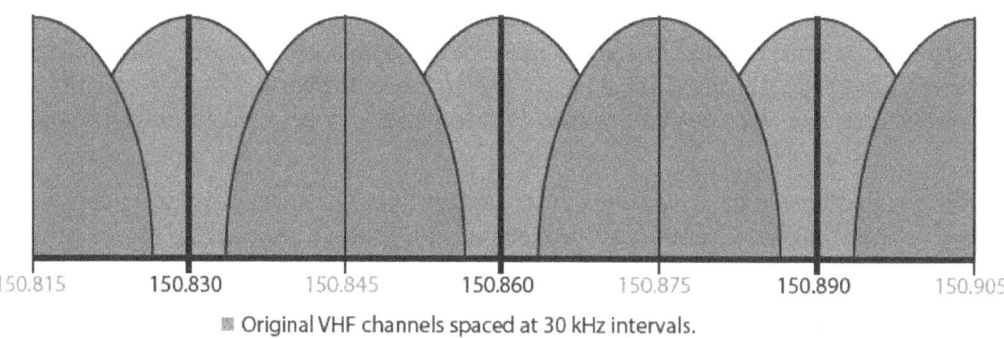

150.815   150.830   150.845   150.860   150.875   150.890   150.905

     ▨ Original VHF channels spaced at 30 kHz intervals.
     ▨ Geographically separated VHF channels spaced
       15 kHz from original VHF channels.

With no unused spectrum available in these bands, the FCC proposed narrowing the bandwidth of the existing frequency assignments, dividing each existing frequency channel in half. Each frequency in the new plan is spaced 7.5 kHz from the previous, and has a bandwidth limited to 12.5 kHz.

Figure 33 – After Narrowbanding.

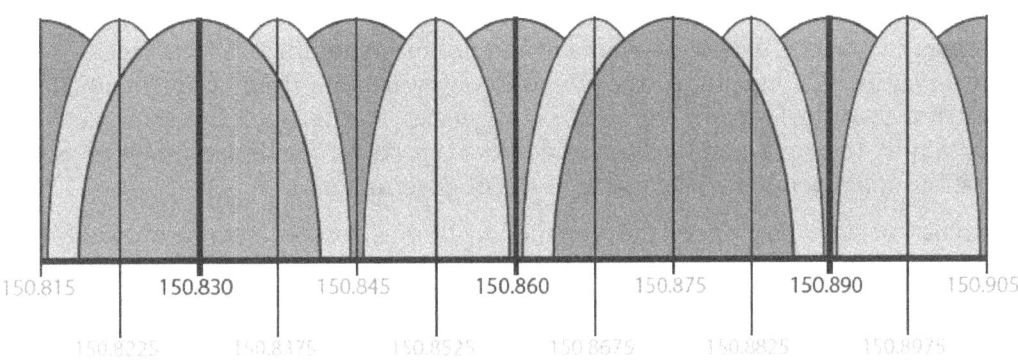

■ Existing 25 kHz bandwidth channels spaced at 30 kHZ intervals.

■ Existing 25 kHz bandwidth channels spaced 15 kHZ from the original channels.

□ New 12.5 kHz bandwidth channels at 7.5 kHZ spacing from existing channels.

The FCC developed a schedule in 1995 for migration from the current band plan to the new narrow band plan. This plan is often called "refarming" to relate it to changing the crops in a field. The schedule for refarming established by the FCC is divided into phases, with each phase increasingly restricting the use of wideband systems to encourage migration to narrowband.

The first phase began in 1997 with the FCC denying certification for equipment that operated with 25 kHz bandwidth if it did not also operate at 12.5 kHz or equivalent bandwidth. This prevented manufacturers from making equipment that would not be able to be used once the future phases came into effect. The FCC predicted that most (wideband) equipment manufactured before this date would become obsolete and unserviceable before the mandatory narrowband deadline.

At the time of this original order, the FCC also made other orders with respect to expansion of existing systems, creation of new systems, and the manufacture and importation of equipment. These orders staggered the restrictions over several years in an attempt to make the transition to narrowband communications less painful to local agencies. Unfortunately the complexity of the rules confused many agencies and, in 2004, before the new rules took effect, the FCC modified the order to have two deadlines, one in 2011 and the other in 2013.

In January 2011, the FCC will no longer accept applications for new systems, modifications to existing systems, or transmitters that operate using a bandwidth greater than 12.5 kHz or equivalent. In addition, the FCC prohibits the manufacture or import of radio equipment that is capable of operating on a bandwidth greater than 12.5 kHz or equivalent.

The final phase begins in January 2013 and prohibits the operation of radios and radio systems that do not comply with the narrowband requirements. All radios, portable, mobile, repeaters, and base stations, that operate in the VHF or UHF bands must be replaced and the systems they operate in must be redesigned before this date.

The FCC's actions to refarm the VHF and UHF bands resulted in perhaps the most confusing set of orders from the FCC concerning public safety communications, resulting in many unnecessary system replacements. These replacements include departments transitioning to systems that do not meet their operational needs and are unnecessarily costly to procure, operate, and maintain. Agencies can keep their existing communications systems that they have been using for years, provided that they modernize the equipment and design by transitioning to 12.5 kHz bandwidth frequencies and equipment prior to 2013.

Public Safety Wireless Advisory Committee

Although the NPSPAC process provided additional frequencies in the 800 MHz band, in the early 1990s the need for more capacity became evident. This increasing need for more frequency spectrum was not limited to non-Federal agencies, as the Federal government had not made modifications to Federal agency needs in many years. In 1993, the FCC established the PSWAC, under direction from Congress, to address the radio frequency spectrum needs of Federal, State, and local agencies over the next 5 years, and over the next 15 years. The goal was to develop a plan to allocate additional spectrum for all users, as well as establish plans for communications interoperability between all levels of government.

The *Final Report of the PSWAC* recommended the allocation of 2.5 MHz of spectrum below 512 MHz for Federal, State, and local public safety interoperability, and the addition of approximately 25 MHz of new spectrum over the next 5 years, and 70 MHz over 15 years for Federal, State, and local public safety use. Although to date only approximately one-third of the new spectrum requested has been allocated for State and local public safety use, this is more than any request in the last 20 years.

700 MHz Spectrum Allocation

As a result of the PSWAC's recommendation that additional spectrum be allocated to public safety, the FCC allocated 24 MHz of new spectrum. This allocation, from 764 MHz through 776 MHz and 794 MHz through 806 MHz, was part of the spectrum previously allocated to TV channels 60 through 69. This spectrum became available for use by public safety through the transition of television stations to digital systems. This portion of spectrum was chosen because it was adjacent to the existing 800 MHz band also used for public safety communications, and radio equipment could be designed easily to operate in both bands.

In 1998, the FCC issued an order that described the rules for the use of the new frequency band, as well as the new band plan. The order split the allocation of frequencies into four basic classes: general-use frequencies, State frequencies, interoperability frequencies, and wideband frequencies. The general-use frequencies could be licensed by both State and local entities, and the allocation and use of the channels would be governed by an FCC-approved regional plan developed by stakeholders in the region. The State frequencies would be licensed to each State, and could be allocated in any manner the State desires. The interoperability frequencies could be licensed by State, local, and, to a limited degree, Federal agencies, and the allocation and use of the frequencies would be governed by a plan produced by a State Interoperability Executive Committee (SIEC) in each State. The wideband channels were intended to provide the ability to develop regional and local high-speed data systems.

800 MHz Reconfiguration

The initial frequency allocations in the 800 MHz band were made available in 1974 by reallocating the frequencies used by television channels 70 through 83. This spectrum was available for use by public safety, business and industrial users, and cellular systems. The FCC allocated 70 channels to public safety and interleaved these with other channels for business and industrial users. Interleaving means that one channel was allocated to business, the next for industrial, and the next for public safety. This repeated, creating an allocation that is layered with public safety sandwiched between other users. Every public safety channel had a non-public-safety system on either side. Later, many of these channels were allocated to Specialized Mobile Radio (SMR) systems, which are private trunked radio systems used by businesses. Figure 34 shows the interleaved frequency allocation, with SMR systems on either side. The 800 MHz NPSPAC band is the block labeled Public Safety to the right of the Upper 200 SMR block.

Figure 34 – 800 MHz Band before Reconfiguration.

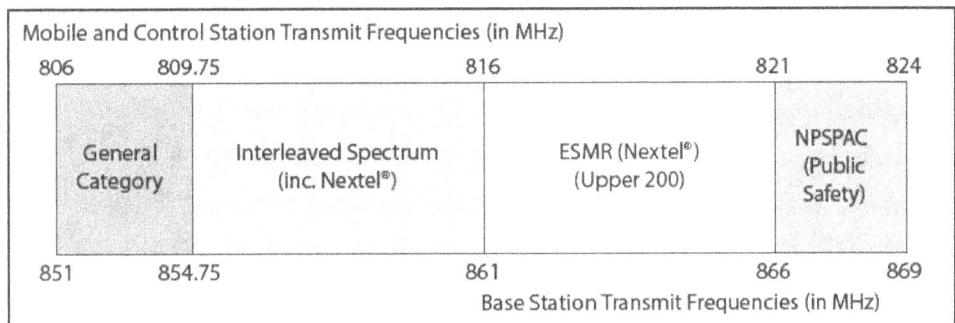

In the early 1990s FleetCall, later to become Nextel®, started to develop a digital SMR network that incorporated the same features as cellular systems. This system used frequencies in the SMR bands, as well as frequencies in the interleaved band. Traditional cellular systems were not allowed to operate in these bands, but FleetCall received waivers from the FCC to operate the new system. At the same time, the deployment of cellular systems was increasing at a rapid pace.

One of the two bands assigned to cellular systems, the Cellular A band is directly adjacent to the NPSPAC 800 MHz band. The NPSPAC band is also adjacent to the Upper SMR band. With FleetCall systems on both sides of the interleaved band, and this and other systems interleaved, along with SMR systems and cellular sandwiching the NPSPAC band, public safety system were in a bad place.

Both the FleetCall system and cellular systems are designed with a large number (30 or more) of transceiver sites throughout the system's coverage area. Compare this with the typical public safety system with one or two sites. The public safety systems were bound to have interference, but none of the system operators that were likely to interfere with the public safety systems recognized the potential.

By the late 1990s, the interference problem with public safety systems in the 800 MHz band had become well-recognized and agencies were demanding action to restore the ability to communicate on emergency incidents. To its credit, the FCC began a process to classify the problem and find a solution. In 2004, the FCC ordered that systems operators in the affected bands must take steps to minimize interference effects. The FCC also ordered that the 800 MHz band would be reconfigured to further minimize the interference from Nextel® and Cellular A band systems under a process known as "rebanding".

Under the rebanding process, Nextel® will fund the effort of relocating existing systems in an equitable manner and in return will receive additional frequencies in the 1.9 MHz band. To supervise the rebanding process, the FCC appointed an independent consulting company BearingPoint as the Transition Administrator (TA). The TA has the responsibility of managing the process, including establishing the schedule, monitoring the process, and facilitating resolution of conflicts. The process was divided into four "waves" that group together the regions that will be reconfigured. All waves were scheduled to be reconfigured by the end of the second quarter of 2008 with the exception of Wave 4, which contains U.S.-Canada or U.S.-Mexico border areas. The reconfiguration of these areas was subject to treaty negotiations that delayed the process.

Figure 35 – 800 MHz Band after Reconfiguration.

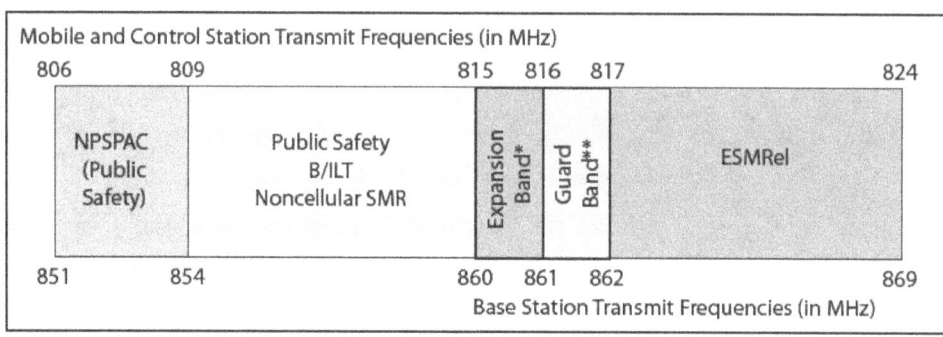

The reconfigured band in the figure above shows that the public safety portions of this new band will be isolated from the ESMR portion of the band where Nextel® operates. In addition, the NPSPAC band has been relocated away from the Cellular A band and is much less likely to suffer significant interference from non-public-safety systems.

Public Safety Broadband Network

In December 2006, the FCC made a statement of opinion in a Notice of Proposed Rulemaking:

> We believe that the time may have come for a significant departure from the typical public safety allocation model the Commission has used in the past…While this system has had significant benefits for public safety users, in terms of permitting them to deploy voice and narrowband facilities for their needs, the system has also resulted in uneven build-out across the country in different bands, balkanization of spectrum between large numbers of incompatible systems, and interoperability difficulties if not inabilities.

This statement predicted the activities that would occur the next April and June, with the FCC's Proposed Rulemaking and Second Report and Order on the 700 MHz band. In this rulemaking, the FCC proposed to create a nationwide public safety broadband data system by rebanding the 700 MHz public safety band to reallocate the 10 MHz wideband frequency allocation and combine this with 10 MHz of new spectrum that would be auctioned. The FCC would allow a single nationwide licensee for the reallocated 10 MHz of existing spectrum, and would auction the other 10 MHz of new spectrum, known as the **D Block**.

The auction would seek to find a bidder that would purchase the rights to the 10 MHz of D Block spectrum and would then have the rights to combine this with the 10 MHz of public safety spectrum to form a nationwide commercial and public safety network. The FCC rules stated that the network must meet the requirements of public safety agencies, and appointed the Public Safety Spectrum Trust Corporation (PSST) to represent the interests of public safety. The PSST developed a Bidder Information Document (BID) that outlines the requirements the new system must meet. These specifications include priority access for public safety users, backup power and networking requirements, and other features necessary to provide a high-reliability system. The PSST also became the licensee for the 10 MHz of reallocated public safety spectrum.

The D Block auction occurred along with other auctions in the first quarter of 2008. There was one bidder for the Block, but the reserve price (minimum bid) set by the FCC was not met, and D Block was not auctioned successfully. After the auction there was some discussion that the requirements for the system set forth in the BID created too much uncertainty as to the cost of constructing the system. This, along with the uncertainty of how many public safety agencies would participate may have led to the unsuccessful auction. The FCC and the PSST are reconsidering the specifications in the BID, and the FCC is evaluating the entire process for improvement prior to another auction attempt. If the nationwide broadband network is built successfully, it will be the first system of its size built specifically for public safety requirements and could serve as an evaluation model for a possible nationwide voice system in the future.

Summary

One problem common to all recommendations to increase the spectrum allocated to public safety agencies is accurately defining "public safety agencies", and another lies in the politics of the allocations. Many State and local governments, and their communications managers in particular, lobbied to include "public service" or "public safety support" agencies in those eligible to license spectrum allocated to public safety.

The result of this is that agencies that do not support the emergency response aspect of public safety are eligible for licenses under the new rules. This includes such diverse groups as school bus companies, road and highway maintenance crews, and public solid-waste disposal agencies. In essence, any State or local government workgroup is eligible for licensing spectrum allocated to public safety, no matter how removed the agency is from emergency response activities. The benefit to State and local governments is that they can build communications systems that support all divisions with spectrum allocated to public safety. Unfortunately this is done by exploiting the public's understanding of what falls under the umbrella of public safety and ultimately reducing the spectrum available for emergency response.

Not all of the responsibility for the lack of adequate spectrum for public safety response lies with the FCC or the various coordinating agencies. Public safety agencies themselves often perpetuate this inadequacy through their actions (or inactions). The insistence by many agencies to maintain "stovepipe" communications systems that duplicate the coverage of other systems and do not operate with neighboring agencies is one of the most egregious examples. The efficiency of frequency use could be increased dramatically if all agencies were committed to cooperative system development with the goal of maximum frequency use among all agencies in a system. The FCC and communications equipment industry are driven by the need to accommodate additional users in a limited amount of radio spectrum and economic forces. The fire service has an opportunity to be a part of the solution to this issue through coordinated organizational participation in the process. If the fire service cannot communicate its needs, or if the fire service voice is fragmented, then a solution will be imposed, and it is unlikely that solution will meet all the operational needs of the service.

SUMMARY

Communications systems for public safety use the same basic communication technologies as other industries, but the needs of the fire service often are unique. These unique requirements, primarily the frequent use in IDLH environments, require different solutions than those of other radio system users. It is important that fire service members communicate these needs when agencies are planning, implementing, and managing their radio systems.

Agencies that are considering remaining on conventional systems in the VHF high or UHF band should be concerned primarily with ensuring their communications systems are safely transitioned to 12.5 kHz bandwidth. In most cases, this will require replacing equipment and redesigning the system to provide comparable coverage. Another consideration is the development of regional communications systems shared among multiple fire departments. This type of system sharing leverages the frequencies available to multiple agencies to develop a system with more surge capacity for all participating agencies.

Departments that already are operating on trunking systems in the 700 MHz or 800 MHz bands should evaluate adding conventional direct-mode channels for fireground use. Over the past 5 to 10 years, this has become the best practice for trunking systems. In addition, agencies should evaluate the coverage and loading of the trunking system continuously to ensure that it still meets the department's needs. This is particularly important in areas that are growing in population, or where significant development is occurring.

If your agency is considering a transition to a trunked 700 MHz or 800 MHz radio system, the most important items to consider are the use of direct-mode communications on the fireground, and concerns about the use of digital voice systems. Any new system must provide for the operational needs of the users in all situations. It is important to participate actively in the specification and implementation of any new communications system.

Interoperability is discussed almost constantly in the public safety communications field. Unfortunately it is often the primary topic, with the day-to-day communications needs ignored. When designing a new or improved communications system, design for the work that happens every day, and consider interoperability as necessary to meet the daily need and special situations.

The fire service has unique communications needs related to operating in hazardous atmospheres with protective equipment. Although the general communications needs of the fire service often are represented, it is important that these needs are presented clearly to the manufacturers, standards-making bodies, and regulatory agencies. The only way to achieve a favorable outcome is to participate and inform.

Endnotes

1. "Star Wars" copyright© Lucasfilm Ltd.

2. Kushner, William M., Michelle Harton, Robert J. Novorita, and Michael J. McLaughlin. "The Acoustic Properties of SCBA Equipment and Effects on Speech Recognition." *IEEE Communications Magazine,* Jan. 2006.

3. Please see www.iafc.org/displayindustryarticle.cfm?articlenbr=33118

4. NIST Technical Note 1477, *Testing of Portable Radios in a Firefighting Environment.*

5. NFPA 1221 is available for purchase from the National Fire Protection Association, www.nfpa.org, (800) 344-3555 or (617) 770-3000.

6. Holloway, Christopher L., Galen Koepke, Dennis Camell, Kate A. Remley, and Dylan Williams. *Radio Propagation Measurements During a Building Collapse: Applications for First Responders.*

7. Phoenix Fire Department Radio System Safety Project (www.phoenix.gov/fire/radioreport.pdf).

8. Phoenix Fire Digital Vehicular Repeater System — Phase 2 Test Report (www.phoenix.gov/fire/).

www.ingramcontent.com/pod-product-compliance
Lightning Source LLC
Chambersburg PA
CBHW081218170526
45165CB00009B/2859